W9-CEH-583

On Becoming a Person

A Therapist's View of Psychotherapy

On Becoming a Person

A Therapist's View of Psychotherapy

Carl R. Rogers

Introduction by
Peter D. Kramer, M.D.

Houghton Mifflin Company
Boston New York

For information about permission to reproduce selections from this book, write to Permissions, Houghton Mifflin Company, 215 Park Avenue South, New York, New York 10003.

For information about this and other Houghton Mifflin trade and reference books and multimedia products, visit The Bookstore at Houghton Mifflin on the World Wide Web at http://www.hmco.com/trade/.

ISBN 0-395-75531-X

Printed in the United States of America

QW/F 15 14

Contents

PART IV

A PHILOSOPHY OF PERSONS

PART V

GETTING AT THE FACTS:

THE PLACE OF RESEARCH IN PSYCHOTHERAPY

PART VI

WHAT ARE THE IMPLICATIONS FOR LIVING?

PART VII

THE BEHAVIORAL SCIENCES AND THE PERSON

Introduction

The publication, in 1961, of *On Becoming a Person* brought Carl Rogers unexpected national recognition. A researcher and clinician, Rogers had believed he was addressing psychotherapists and only after the fact discovered that he "was writing for *people*—nurses, housewives, people in the business world, priests, ministers, teachers, youth." The book sold millions of copies when million was a rare number in publishing. Rogers was, for the decade that followed, the Psychologist of America, likely to be consulted by the press on any issue that concerned the mind, from creativity to self-knowledge to the national character.

Certain ideas that Rogers championed have become so widely accepted that it is difficult to recall how fresh, even revolutionary, they were in their time. Freudian psychoanalysis, the prevailing model of mind at mid-century, held that human drives—sex and aggression—were inherently selfish, constrained at a price and with difficulty by the forces of culture. Cure, in the Freudian model, came through a relationship that frustrated the patient, fostering anxiety necessary for the patient to accept the analyst's difficult truths. Rogers, in contrast, believed that people need a relationship in which they are accepted. The skills the Rogerian therapist uses are empathy—a word that in Freud's time was largely restricted to the feelings with which an observer invests a work of art—and "unconditional positive regard." Rogers stated his central hypothesis in one sentence: "If I can provide a certain type of relationship, the other will discover within himself the capacity to use that relationship for growth, and change and personal development will occur." By growth, Rogers meant movement in the direction of self-esteem, flexibility, respect for self and others. To Rogers, man is "incorrigibly socialized in his desires." Or, as Rogers puts the matter repeatedly, when man is most fully man, he is to be trusted.

Rogers was, in Isaiah Berlin's classification, a hedgehog: He knew one thing, but he knew it so well that he could make a world of it. From Rogers comes our contemporary emphasis on self-esteem and its power to mobilize a person's other strengths. Rogers's understanding of acceptance as the ultimate liberating force implies that people who are not ill can benefit from therapy and that nonprofessionals can act as therapists; the modern self-help group arises quite directly from Rogers's human potential movement. That marriage, like therapy, depends on genuineness and empathy is basic Carl Rogers. It is Rogers, much more than Benjamin Spock, who speaks for nondirective parenting and teaching.

It is ironic that while Rogers's ideas are in the ascendant—so much so that they are now attacked as powerful cultural assumptions in need of revision—his writings are in eclipse. This is a shame, because a culture should know where its beliefs originate and because Rogers's writing remains lucid, charming, and accessible.

Certainly Rogers's ideas prevail within the mental health professions. Today's cutting-edge school of psychoanalysis is called "self psychology," a name Rogers could have coined. Like client-centered therapy, which Rogers developed in the 1940s, self psychology understands relationship, more than insight, to be central to change; and like client-centered psychotherapy, self psychology holds that the optimal level of frustration is "as little as possible." The therapeutic posture in self psychology resembles nothing so much as unconditional positive regard. But self psychology—founded in Chicago, when Rogers was a preeminent figure there—has given Rogers nary a word of credit.

Much of the explanation has to do with who Rogers was. American rather than European, farm-raised rather than urban (he was born in Chicago but moved to the country at age twelve and said his respect for the experimental method arose from his reading, in adolescence, of a long text called *Feeds and Feeding*), midwestern rather than eastern, sanguine rather than melancholic, accessible and open, Rogers displayed none of the dark complexity of the postwar intellectual. Rogers's openness—in an important sense *On Becoming a Person* needs no introduction, since Rogers introduces himself in an essay exactly titled "This is

Me"—stands in contrast to the posture favored by his peers, who believed the therapist must present himself as a blank slate. The prevailing judgment was that Rogers could be dismissed because he was not serious.

This judgment hides and reveals a narrow view of what is serious or intellectual. Rogers was a university professor and a widely published scholar, with sixteen books and more than two hundred articles to his credit. The very success of *On Becoming a Person* may have injured Rogers's academic reputation; he was known for the directness and simplicity of these essays, not for the complexity of more technical theoretical articles written in the same period. But even in *On Becoming a Person*, Rogers places his ideas in historical and social context, alluding to contemporary social psychology, animal ethology, and communications and general systems theory. He locates his heritage in existential philosophy, referring most often to Søren Kierkegaard (from whom he takes the phrase "to be that self which one truly is," Rogers's answer to the question "What is the goal of life?") and Martin Buber. And Rogers enjoyed a busy career as a public intellectual, debating and corresponding openly with such figures as Buber, Paul Tillich, Michael Polanyi, Gregory Bateson, Hans Hofman, and Rollo May.

More than most of his colleagues, Rogers was a committed scientist espousing an empirical evaluation of psychotherapy. As early as the 1940s, and before anyone else in the field, Rogers was recording psychotherapeutic sessions for the purpose of research. He is the first inventor of a psychotherapy to define his approach in operational terms, listing six necessary and sufficient conditions (engaged patient, empathic therapist, etc.) for constructive personality change. He developed reliable measures and sponsored and publicized evaluations of his hypotheses. Rogers was committed to an assessment of process: What helps people to change? His research, and that of his scientific collaborators, led to results embarrassing to the psychoanalytic establishment. For example, one study, of transcripts of therapy sessions, found that in response to clarification and interpretation—the tools of psychoanalysis—clients typically abandon self-exploration; only reflection of feeling by the therapist leads directly to further exploration and new insight.

Rogers, in other words, marshaled a substantial intellectual effort in the service of a simple belief: Humans require acceptance, and given acceptance, they move toward "self-actualization." The corollaries of this hypothesis were evident to Rogers and his contemporaries. The complex edifice of psychoanalysis is unnecessary—transference may well exist, but to explore it is unproductive. A haughty and distant posture, the one assumed by many psychoanalysts at mid-century, is certainly countertherapeutic. The self-awareness and human presence of the therapist is more important than the therapist's technical training. And the boundary between psychotherapy and ordinary life is necessarily thin. If acceptance, empathy, and positive regard are the necessary and sufficient conditions for human growth, then they ought equally to inform teaching, friendship, and family life.

These ideas offended a number of establishments—psychoanalytic, educational, religious. But they were welcome to a broad segment of the public. They informed the popular dialogue of the 1960s—many of the on-campus demands of sixties protesters relied implicitly on Rogers's beliefs about human nature—and they helped shape our institutions for the remainder of the century.

Before being dismissed and forgotten, Rogers was attacked on a series of particular grounds. Reviews of the research literature showed the necessity and sufficiency of his six conditions difficult to prove, although the evidence favoring a present, empathic stance on the part of the therapist remains strong. Rogers's notion that therapist and client can meet on equal ground was challenged early on by Martin Buber and more recently by a contentious critic of psychotherapy, Jeffrey Masson. (In a lovely little book titled simply *Carl Rogers* [London: Sage Publications, 1992], Brian Thorne reviews and, with some success, rebuts these criticisms.) As our distance from Rogers grows, the critiques seem increasingly irrelevant. What Rogers provides—what all great therapists provide—is a unique vision.

It is clear that the mid-century psychoanalytic theory of man was incomplete. Freud and, more starkly, Melanie Klein, the founder of a school of psychoanalysis that has had enormous influence over modern views of intense human relationships, captured humanity's dark side, that part of our animal heritage that includes the violence and competitive sexuality related to struggles for hierarchy dominance. They ignored a reproductive strategy that coexists with hierarchy dominance and is also strongly encoded in genes and culture: reciprocity and altruism. Animal ethologists and evolutionary biologists today would agree with Rogers's thesis that when a human being is adequately accepted, it is these latter traits that are likely to predominate.

Buber—not only as a religious philosopher but as a student of Eugen Bleuler, the great German descriptive psychiatrist—was doubtless justified in his skepticism over Rogers's contention that man, ill or well, is to be trusted. But Freud, Klein, and Buber were thoroughly enmeshed in Old World perspectives. Rogers's relentless optimism is perhaps best seen as one of many interesting attempts to bring to psychotherapy the flavor of the New World.

In this endeavor Rogers had many peers. Harry Stack Sullivan added a number of facets to psychoanalysis: attention to the influence of the chum in childhood development; exploration of the patient's particular social environment; and active use of the therapist's self to block patients' characteristic projections. Murray Bowen turned attention from the patient's family in childhood (the Oedipus constellation) to the present family, and he freed the therapist to act as a sort of coach in the patient's effort to find room within the family's rigid structure. Milton Erickson revived hypnotic techniques and used them impishly, turning the therapist into a master manipulator who catapults the patient past developmental impasses. Carl Whitaker stressed the hindrance of theory in clinical practice, demanding of the therapist both an existential presence and an awareness of local family customs. To this list could be added the names of immigrants—Erich Fromm, Victor Frankl, Hellmuth Kaiser, Erik Erikson, Heinz Kohut—whose work took on a decidedly American cast, free and experimental and socially aware.

Although he rejects the Puritan premise of original sin, Rogers—in taking care to understand the other as a free individual, in focusing on his own authenticity and active presence, in trusting the positive potential in each client—creates a therapeutic view of man that conforms to important aspects of the American ethos. Rogers's central premise is that people are inherently resourceful. For Rogers, the cardinal sin in therapy, or in teaching or family life, is the imposition of authority. A radical egalitarian, Rogers sees individuals as capable of self-direction without regard for received wisdom and outside of organizations such as the church or the academy. Despite its origins in the helping relationship, Rogers's philosophy is grounded in Thoreau and Emerson, in the primacy of self-reliance.

In embracing Rogers, Americans took important parts of themselves to heart—parts about which, however, the nation remains ambivalent. Does individualism imply fresh exploration of values by each person in each new generation, or must individualism be linked to fixed traditions and a view of man as selfish and competitive? Returning to established curricula and orthodox values, conservatives today attack not only Rogers but also an important strain of American humanism. It is perhaps because of Rogers's American core that he is so much more respected—understood as a distinctive voice, taught with earnestness—in dozens of countries outside the United States.

Rogers's voice—warm, enthusiastic, confident, concerned—is what binds the disparate essays in *On Becoming a Person*. We encounter a man trying patiently, but with all the resources at his command, to hear others and himself. This attentive listening is in the service of both the individual and the grand question, what it means to become a person. In describing clients, Rogers assumes the language and prosody of existentialism. Of one struggling man, Rogers writes, "At that moment he is nothing but his pleadingness, all the way through . . . [F]or that moment he *is* his dependency, in a way which astonishes him."

Any notion that Rogers is not serious, not aware of human frailty, not intellectual must dissolve in response to his transcripts of painstaking clinical work. Rogers does what generations of psychology students have satirized him for doing, namely, repeat clients' words. But he also summarizes clients' feelings with precision, beauty of expression, and generous tentativeness. And he has a genius for accepting others.

In her fifth psychotherapy session with Rogers, Mrs. Oak, a troubled homemaker, catches herself singing a "sort of a song without any music." Rogers's summary of her sequence of feelings leads Mrs. Oak to amplify inner experiences and explore her metaphor. We hear a person grasping for elusive authenticity, denigrating her own thoughts: "And then there just seems to be this flow of words which somehow aren't forced and then occasionally this doubt creeps in. Well, it sort of takes form of a, maybe you're just making music." Like all humans, in Rogers's schema, Mrs. Oak begins as remote from the self; with acceptance, she will remove façades and achieve actualization. In her ninth session, Mrs. Oak reveals, in embarrassed fashion, a limited form of self-confidence: ". . . I have had what I have come to call to myself, told myself were 'flashes of sanity'. . . It's just a feeling once in a while of finding myself a whole kind of person in a terribly chaotic kind of world." But she cannot reveal this confident self to others. Rogers immediately recalls the earlier session: "A feeling that it wouldn't be *safe* to talk about the singing you . . . Almost as if there was no place for such a person to, to exist." Such attunement to the other is high art, though it is hard to know whether Rogers is capturing the client's inner melody or supplying one of his own composing.

This ambiguity remains regarding Rogers's clinical work: Did he merely, as he claimed, accept the other, or did he provide parts of his own well-differentiated self? What is unambiguous, as we read Rogers today, is his extensive contribution to contemporary culture, to our sense of who we are. It is a pleasure to encounter him again, to have access once more to his music.

PETER D. KRAMER, M.D.

To the Reader

THOUGH IT SHOCKS ME SOMEWHAT TO SAY SO, I have been a psychotherapist (or personal counselor) for more than thirty-three years. This means that during a period of a third of a century I have been trying to be of help to a broad sampling of our population: to children, adolescents and adults; to those with educational, vocational, personal and marital problems; to "normal," "neurotic," and "psychotic" individuals (the quotes indicate that for me these are all misleading labels); to individuals who come for help, and those who are sent for help; to those whose problems are minor, and to those whose lives have become utterly desperate and without hope. I regard it as a deep privilege to have had the opportunity to know such a diverse multitude of people so personally and intimately.

Out of the clinical experience and research of these years I have written several books and many articles. The papers in this volume are selected from those I have written during the most recent ten of the thirty-three years, from 1951 to 1961. I would like to explain the reasons that I have for gathering them into a book.

In the first place I believe that almost all of them have relevance for personal living in this perplexing modern world. This is in no sense a book of advice, nor does it in any way resemble the "do-it-yourself" treatise, but it has been my experience that readers of these papers have often found them challenging and enriching. They have to some small degree given the person more security in making and following his personal choices as he endeavors to move toward being the person he would like to be. So for this reason I should like to have them more widely available to any who might be interested—to "the intelligent layman," as the phrase goes. I feel this especially since all of my previous books have been published for the professional psychological audience, and have never been readily available to the person outside of that group. It is my sincere hope that many people who have no particular interest in the field of counseling or psychotherapy will find that the learnings emerging in this field will strengthen them in their own living. It is also my hope and belief that many people who have never sought counseling help will find, as they read the excerpts from the recorded therapy interviews of the many clients in these pages, that they are subtly enriched in courage and self confidence, and that understanding of their own difficulties will become easier as they live through, in their imagination and feeling, the struggles of others toward growth.

Another influence which has caused me to prepare this book is the increasing number and urgency of requests from those who are already acquainted with my point of view in counseling, psychotherapy, and interpersonal relationships. They have made it known that they wish to be able to obtain accounts of my more recent thinking and work in a convenient and available package. They are frustrated by hearing of unpublished articles which they cannot acquire; by stumbling across papers of mine in out-of-the-way journals; they want them brought together. This is a flattering request for any author. It also constitutes an obligation which I have tried to fulfill. I hope that they will be pleased with the selection I have made. Thus in this respect this volume is for those psychologists, psychiatrists, teachers, educators, school counselors, religious workers, social workers, speech therapists, industrial leaders, labor-management specialists, political scientists and others who have in the past found my work relevant to their professional efforts. In a very real sense, it is dedicated to them.

There is another motive which has impelled me, a more complex and personal one. This is the search for a suitable audience for what I have to say. For more than a decade this problem has puzzled me. I know that I speak to only a fraction of psychologists. The majority—their interests suggested by such terms as stimulus-response, learning theory, operant conditioning—are so committed to seeing the individual solely as an object, that what I have to say often baffles if it does not annoy them. I also know that I speak to but a fraction of psychiatrists. For many, perhaps most of them, the truth about psychotherapy has already been voiced long ago by Freud, and they are uninterested in new possibilities, and uninterested in or antagonistic to research in this field. I also know that I speak to but a portion of the divergent group which call themselves counselors. The bulk of this group are primarily interested in predictive tests and measurements, and in methods of guidance.

So when it comes to the publication of a particular paper, I have felt dissatisfied with presenting it to a professional journal in any one of these fields. I have published articles in journals of each of these types, but the majority of my writings in recent years have piled up as unpublished manuscripts, distributed privately in mimeographed form. They symbolize my uncertainty as to how to reach whatever audience it is I am addressing.

During this period journal editors, often of small or highly specialized journals, have learned of some of these papers, and have requested permission to publish. I have always acceded to these requests, with the proviso that I might wish to publish the paper elsewhere at some later time. Thus the majority of the papers I have written during this decade have been unpublished, or have seen the light of day in some small, or specialized, or off-beat journal.

Now however I have concluded that I wish to put these thoughts out in book form so that they can seek their *own* audience. I am sure that that audience will cut across a variety of disciplines, some of them as far removed from my own field as philosophy and the science of government. Yet I have come to believe that the audience will have a certain unity, too. I believe these papers belong in a trend which is having and will have its impact on psychology, psychiatry, philosophy, and other fields. I hesitate to label such a trend but in my mind there are associated with it adjectives such as phenomenological, existential, person-centered; concepts such as self-actualization, becoming, growth; individuals (in this country) such as Gordon Allport, Abraham Maslow, Rollo May. Hence, though the group to which this book speaks meaningfully will, I believe, come from many disciplines, and have many wide-ranging interests, a common thread may well be their concern about the person and his becoming, in a modern world which appears intent upon ignoring or diminishing him.

There is one final reason for putting out this book, a motive which means a great deal to me. It has to do with the great, in fact the desperate, need of our times for more basic knowledge and more competent skills in dealing with the tensions in human relationships. Man's awesome scientific advances into the infinitude of space as well as the infinitude of sub-atomic particles seems most likely to lead to the total destruction of our world unless we can make great advances in understanding and dealing with interpersonal and inter-group tensions. I feel very humble about the modest knowledge which has been gained in this field. I hope for the day when we will invest at least the price of one or two large rockets in the search for more adequate understanding of human relationships. But I also feel keenly concerned that the knowledge we *have* gained is very little recognized and little utilized. I hope it may be clear from this volume that we *already* possess learnings which, put to use, would help to decrease the inter-racial, industrial, and international tensions which exist. I hope it will be evident that these learnings, used preventively, could aid in the development of mature, nondefensive, understanding persons who would deal constructively with future tensions as they arise. If I can thus make clear to a significant number of people the unused resource knowledge already available in the realm of interpersonal relationships, I will feel greatly rewarded.

So much for my reasons for putting forth this book. Let me conclude with a few comments as to its nature. The papers which are brought together here represent the major areas of my interest during the past decade.* They were prepared for different purposes, usually for different audiences, or formulated simply for my own satisfaction. I have written for each chapter an introductory

* The one partial exception is in the area of explicit theory of personality. Having just recently published a complete and technical presentation of my theories in a book which should be available in any professional library, I have not tried to include such material here. The reference referred to is my chapter entitled, "A theory of therapy, personality, and interpersonal relationships as developed in the client-centered framework" in Koch, S. (ed.) *Psychology: A Study of a Science*, vol. III, pp. 184–256. McGraw-Hill, 1959.

note which tries to set the material in an understandable context. I have organized the papers in such a way that they portray a unified and developing theme from the highly personal to the larger social significance. In editing them, I have eliminated duplication, but where different papers present the same concept in different ways I have often retained these "variations on a theme" hoping that they might serve the same purpose as in music, namely to enrich the meaning of the melody. Because of their origin as separate papers, each one can be read independently of the others if the reader so desires.

Stated in the simplest way, the purpose of this book is to share with you something of my experience—something of me. Here is what I have experienced in the jungles of modern life, in the largely unmapped territory of personal relationships. Here is what I have seen. Here is what I have come to believe. Here are the ways I have tried to check and test my beliefs. Here are some of the perplexities, questions, concerns and uncertainties which I face. I hope that out of this sharing you may find something which speaks to you.

Departments of Psychology and Psychiatry
The University of Wisconsin
April, 1961

PART I

Speaking Personally

I speak as a person, from a context of personal experience and personal learnings.

1

"This is Me"

The Development of My Professional Thinking and Personal Philosophy

This chapter combines two very personal talks. Five years ago I was asked to speak to the senior class at Brandeis University to present, not my ideas of psychotherapy, but myself. How had I come to think the thoughts I had? How had I come to be the person I am? I found this a very thought-provoking invitation, and I endeavored to meet the request of these students. During this past year the Student Union Forum Committee at Wisconsin made a somewhat similar request. They asked me to speak in a personal vein on their "Last Lecture" series, in which it is assumed that, for reasons unspecified, the professor is giving his last lecture and therefore giving quite personally of himself. (It is an intriguing comment on our educational system that it is assumed that only under the most dire circumstances would a professor reveal himself in any personal way.) In this Wisconsin talk I expressed more fully than in the first one the personal learnings or philosophical themes which have come to have meaning for me. In the current chapter I have woven together both of these talks, trying to retain something of the informal character which they had in their initial presentation.

The response to each of these talks has made me realize how hun-

gry people are to know something of the person *who is speaking to them or teaching them. Consequently I have set this chapter first in the book in the hope that it will convey something of me, and thus give more context and meaning to the chapters which follow.*

I HAVE BEEN INFORMED that what I am expected to do in speaking to this group is to assume that my topic is "This is Me." I feel various reactions to such an invitation, but one that I would like to mention is that I feel honored and flattered that any group wants, in a personal sense, to know who I am. I can assure you it is a unique and challenging sort of invitation, and I shall try to give to this honest question as honest an answer as I can.

So, who am I? I am a psychologist whose primary interest, for many years, has been in psychotherapy. What does that mean? I don't intend to bore you with a long account of my work, but I would like to take a few paragraphs from the preface to my book, *Client-Centered Therapy*, to indicate in a subjective way what it means to me. I was trying to give the reader some feeling for the subject matter of the volume, and I wrote as follows. "What is this book about? Let me try to give an answer which may, to some degree, convey the living experience that this book is intended to be.

"This book is about the suffering and the hope, the anxiety and the satisfaction, with which each therapist's counseling room is filled. It is about the uniqueness of the relationship each therapist forms with each client, and equally about the common elements which we discover in all these relationships. This book is about the highly personal experiences of each one of us. It is about a client in my office who sits there by the corner of the desk, struggling to be himself, yet deathly afraid of being himself — striving to see his experience as it is, wanting to *be* that experience, and yet deeply fearful of the prospect. This book is about me, as I sit there with that client, facing him, participating in that struggle as deeply and sensitively as I am able. It is about me as I try to perceive his experience, and the meaning and the feeling and the taste and the flavor that it has for him. It is about me as I bemoan my very human fallibility in understanding that client, and the occasional failures to see life as it appears to him, failures which fall like heavy objects across the intricate, delicate web of growth which is taking place. It is about me as I rejoice at the privilege of being a midwife to a new personality — as I stand by with awe at the emergence of a self, a person, as I see a birth process in which I have had an important and facilitating part. It is about both the client and me as we regard with wonder the potent and orderly forces which are evident in this whole experience, forces which seem deeply rooted in the universe as a whole. The book is, I believe, about life, as life vividly reveals itself in the therapeutic process — with its blind power and its tremendous capacity for destruction, but with its overbalancing thrust toward growth, if the opportunity for growth is provided."

Perhaps that will give you some picture of what I do and the way I feel about it. I presume you may also wonder how I came to engage in that occupation, and some of the decisions and choices, conscious and unconscious, which were made along the way. Let me see if I can give you some of the psychological highlights of my autobiography, particularly as it seems to relate to my professional life.

My Early Years

I was brought up in a home marked by close family ties, a very strict and uncompromising religious and ethical atmosphere, and what amounted to a worship of the virtue of hard work. I came along as the fourth of six children. My parents cared a great deal for us, and had our welfare almost constantly in mind. They were also, in many subtle and affectionate ways, very controlling of our behavior. It was assumed by them and accepted by me that we were different from other people — no alcoholic beverages, no dancing, cards or theater, very little social life, and *much* work. I have a hard time convincing my children that even carbonated beverages had a faintly sinful aroma, and I remember my slight feeling of wickedness when I had my first bottle of "pop." We had good times together within the family, but we did not mix. So I was a pretty solitary boy, who read incessantly, and went all through high school with only two dates.

When I was twelve my parents bought a farm and we made our home there. The reasons were twofold. My father, having become a prosperous business man, wanted it for a hobby. More important, I believe, was the fact that it seemed to my parents that a growing adolescent family should be removed from the "temptations" of suburban life.

Here I developed two interests which have probably had some real bearing on my later work. I became fascinated by the great night-flying moths (Gene Stratton-Porter's books were then in vogue) and I became an authority on the gorgeous Luna, Polyphemus, Cecropia and other moths which inhabited our woods. I laboriously bred the moths in captivity, reared the caterpillars, kept the cocoons over the long winter months, and in general realized some of the joys and frustrations of the scientist as he tries to observe nature.

My father was determined to operate his new farm on a scientific basis, so he bought many books on scientific agriculture. He encouraged his boys to have independent and profitable ventures of our own, so my brothers and I had a flock of chickens, and at one time or other reared from infancy lambs, pigs and calves. In doing this I became a student of scientific agriculture, and have only realized in recent years what a fundamental feeling for science I gained in that way. There was no one to tell me that Morison's *Feeds and Feeding* was not a book for a fourteen-year-old, so I ploughed through its hundreds of pages, learning how experiments were conducted — how control groups were matched with experimental groups, how conditions were held constant by randomizing procedures, so that the influence of a given food on meat production or milk production could be established. I learned how difficult it is to test an hypothesis. I acquired a knowledge of and a respect for the methods of science in a field of practical endeavor.

College and Graduate Education

I started in college at Wisconsin in the field of agriculture. One of the things I remember best was the vehement statement of an agronomy professor in regard to the learning and use of facts. He stressed the futility of an encyclopedic knowledge for its own sake, and wound up with the injunction, "Don't be a damned ammunition wagon; be a rifle!"

During my first two college years my professional goal changed, as the result of some emotionally charged student religious conferences, from that of a scientific agriculturist to that of the ministry — a slight shift! I changed from agriculture to history, believing this would be a better preparation.

In my junior year I was selected as one of a dozen students from this country to go to China for an international World Student Christian Federation Conference. This was a most important experience for me. It was 1922, four years after the close of World War I. I saw how bitterly the French and Germans still hated each other, even though as individuals they seemed very likable. I was forced to stretch my thinking, to realize that sincere and honest people could believe in very divergent religious doctrines. In major ways I for the first time emancipated myself from the religious thinking of my parents, and realized that I could not go along with them. This independence of thought caused great pain and stress in our relationship, but looking back on it I believe that here, more than at any other one time, I became an independent person. Of course there was much revolt and rebellion in my attitude during that period, but the essential split was achieved during the six months I was on this trip to the Orient, and hence was thought through away from the influence of home.

Although this is an account of elements which influenced my professional development rather than my personal growth, I wish to mention very briefly one profoundly important factor in my personal life. It was at about the time of my trip to China that I fell in love with a lovely girl whom I had known for many years, even in childhood, and we were married, with the very reluctant consent of our parents, as soon as I finished college, in order that we could go to graduate school together. I cannot be very objective about this, but her steady and sustaining love and companionship during all the years since has been a most important and enriching factor in my life.

I chose to go to Union Theological Seminary, the most liberal in the country at that time (1924), to prepare for religious work. I have never regretted the two years there. I came in contact with some great scholars and teachers, notably Dr. A. C. McGiffert, who believed devoutly in freedom of inquiry, and in following the truth no matter where it led.

Knowing universities and graduate schools as I do now — knowing their rules and their rigidities — I am truly astonished at one very significant experience at Union. A group of us felt that ideas were being fed to us, whereas we wished primarily to explore our own questions and doubts, and find out where they led. We petitioned the administration that we be allowed to set up a seminar for credit, a seminar with no instructor, where the curriculum would be composed of our own questions. The seminary was understandably perplexed by this, but they granted our petition! The only restriction was that in the interests of the institution a young instructor was to sit in on the seminar, but would take no part in it unless we wished him to be active.

I suppose it is unnecessary to add that this seminar was deeply satisfying and clarifying. I feel that it moved me a long way toward a philosophy of life which was my own. The majority of the members of that group, in thinking their way through the questions they had raised, thought themselves right out of religious work. I was one. I felt that questions as to the meaning of life, and the possibility of the constructive improvement of life for individuals, would probably always interest me, but I could not work in a field where I would be required to believe in some specified religious doctrine. My beliefs had already changed tremendously, and might continue to change. It seemed to me it would be a horrible thing to *have* to profess a set of beliefs, in order to remain in one's profession. I wanted to find a field in which I could be sure my freedom of thought would not be limited.

Becoming a Psychologist

But what field? I had been attracted, at Union, by the courses and lectures on psychological and psychiatric work, which were then beginning to develop. Goodwin Watson, Harrison Elliott,

Marian Kenworthy all contributed to this interest. I began to take more courses at Teachers' College, Columbia University, across the street from Union Seminary. I took work in philosophy of education with William H. Kilpatrick, and found him a great teacher. I began practical clinical work with children under Leta Hollingworth, a sensitive and practical person. I found myself drawn to child guidance work, so that gradually, with very little painful readjustment, I shifted over into the field of child guidance, and began to think of myself as a clinical psychologist. It was a step I eased into, with relatively little clearcut conscious choice, rather just following the activities which interested me.

While I was at Teachers' College I applied for, and was granted a fellowship or internship at the then new Institute for Child Guidance, sponsored by the Commonwealth Fund. I have often been grateful that I was there during the first year. The organization was in a chaotic beginning state, but this meant that one could do what he wanted to do. I soaked up the dynamic Freudian views of the staff, which included David Levy and Lawson Lowrey, and found them in great conflict with the rigorous, scientific, coldly objective, statistical point of view then prevalent at Teachers' College. Looking back, I believe the necessity of resolving that conflict in me was a most valuable learning experience. At the time I felt I was functioning in two completely different worlds, "and never the twain shall meet."

By the end of this internship it was highly important to me that I obtain a job to support my growing family, even though my doctorate was not completed. Positions were not plentiful, and I remember the relief and exhilaration I felt when I found one. I was employed as psychologist in the Child Study Department of the Society for the Prevention of Cruelty to Children, in Rochester, New York. There were three psychologists in this department, and my salary was $2,900 per year.

I look back at the acceptance of that position with amusement and some amazement. The reason I was so pleased was that it was a chance to do the work I wanted to do. That, by any reasonable criterion it was a dead-end street professionally, that I would be isolated from professional contacts, that the salary was not good even by the standards of that day, seems not to have occurred to me, as nearly as I can recall. I think I have always had a feeling that if I was given some opportunity to do the thing I was most interested in doing, everything else would somehow take care of itself.

THE ROCHESTER YEARS

The next twelve years in Rochester were exceedingly valuable ones. For at least the first eight of these years, I was completely immersed in carrying on practical psychological service, diagnosing and planning for the delinquent and underprivileged children who were sent to us by the courts and agencies, and in many instances carrying on "treatment interviews." It was a period of relative professional isolation, where my only concern was in trying to be more effective with our clients. We had to live with our failures as well as our successes, so that we were forced to learn. There was only one criterion in regard to any method of dealing with these children and their parents, and that was, "Does it work? Is it effective?" I found I began increasingly to formulate my own views out of my everyday working experience.

Three significant illustrations come to mind, all small, but important to me at the time. It strikes me that they are all instances of disillusionment — with an authority, with materials, with myself.

In my training I had been fascinated by Dr. William Healy's writings, indicating that delinquency was often based upon sexual conflict, and that if this conflict was uncovered, the delinquency ceased. In my first or second year at Rochester I worked very hard with a youthful pyromaniac who had an unaccountable impulse to set fires. Interviewing him day after day in the detention home, I gradually traced back his desire to a sexual impulse regarding masturbation. Eureka! The case was solved. However, when placed on probation, he again got into the same difficulty.

I remember the jolt I felt. Healy might be wrong. Perhaps I was learning something Healy didn't know. Somehow this incident impressed me with the possibility that there were mistakes in authoritative teachings, and that there was still new knowledge to discover.

The second naive discovery was of a different sort. Soon after coming to Rochester I led a discussion group on interviewing. I discovered a published account of an interview with a parent, approximately verbatim, in which the case worker was shrewd, insightful, clever, and led the interview quite quickly to the heart of the difficulty. I was happy to use it as an illustration of good interviewing technique.

Several years later, I had a similar assignment and remembered this excellent material. I hunted it up again and re-read it. I was appalled. Now it seemed to me to be a clever legalistic type of questioning by the interviewer which convicted this parent of her unconscious motives, and wrung from her an admission of her guilt. I now knew from my experience that such an interview would not be of any lasting help to the parent or the child. It made me realize that I was moving away from any approach which was coercive or pushing in clinical relationships, not for philosophical reasons, but because such approaches were never more than superficially effective.

The third incident occurred several years later. I had learned to be more subtle and patient in interpreting a client's behavior to him, attempting to time it in a gentle fashion which would gain acceptance. I had been working with a highly intelligent mother whose boy was something of a hellion. The problem was clearly her early rejection of the boy, but over many interviews I could not help her to this insight. I drew her out, I gently pulled together the evidence she had given, trying to help her see the pattern. But we got nowhere. Finally I gave up. I told her that it seemed we had both tried, but we had failed, and that we might as well give up our contacts. She agreed. So we concluded the interview, shook hands, and she walked to the door of the office. Then she turned and asked, "Do you ever take adults for counseling here?" When I replied in the affirmative, she said, "Well then, I would like some help." She came to the chair she had left, and began to pour out her despair about her marriage, her troubled relationship with her husband, her sense of failure and confusion, all very different from the sterile "case history" she had given before. Real therapy began then, and ultimately it was very successful.

This incident was one of a number which helped me to experience the fact — only fully realized later — that it is the *client* who knows what hurts, what directions to go, what problems are crucial, what experiences have been deeply buried. It began to occur to me that unless I had a need to demonstrate my own cleverness and learning, I would do better to rely upon the client for the direction of movement in the process.

PSYCHOLOGIST OR ?

During this period I began to doubt that I was a psychologist. The University of Rochester made it clear that the work I was doing was not psychology, and they had no interest in my teaching in the Psychology Department. I went to meetings of the American Psychological Association and found them full of papers on the learning processes of rats and laboratory experiments which seemed to me to have no relation to what I was doing. The psychiatric social workers, however, seemed to be talking my language, so I became active in the social work profession, moving up to local and even national offices. Only when the American Association for Applied Psychology was formed did I become really active as a psychologist.

I began to teach courses at the University on how to understand and deal with problem children, under the Department of Sociology. Soon the Department of Education wanted to classify these as education courses, also. [Before I left Rochester, the Department of Psychology, too, finally requested permission to list them, thus at last accepting me as a psychologist.] Simply describing these experiences makes me realize how stubbornly I have followed my own course, being relatively unconcerned with the question of whether I was going with my group or not.

Time does not permit to tell of the work of establishing a separate Guidance Center in Rochester, nor the battle with some of the psychiatric profession which was included. These were largely administrative struggles which did not have too much to do with the development of my ideas.

MY CHILDREN

It was during these Rochester years that my son and daughter grew through infancy and childhood, teaching me far more about individuals, their development, and their relationships, than I could ever have learned professionally. I don't feel I was a very good parent in their early years, but fortunately my wife was, and as time went on I believe *I* gradually became a better and more understanding parent. Certainly the privilege during these years and later, of being in relationship with two fine sensitive youngsters through all their childhood pleasure and pain, their adolescent assertiveness and difficulties, and on into their adult years and the beginning of their own families, has been a priceless one. I think my wife and I regard as one of the most satisfying achievements in which we have had a part, the fact that we can really communicate in a deep way with our grown-up children and their spouses, and they with us.

Ohio State Years

In 1940 I accepted a position at Ohio State University. I am sure the only reason I was considered was my book on the *Clinical Treatment of the Problem Child,* which I had squeezed out of vacations, and brief leaves of absence. To my surprise, and contrary to my expectation, they offered me a full professorship. I heartily recommend starting in the academic world at this level. I have often been grateful that I have never had to live through the frequently degrading competitive process of step-by-step promotion in university faculties, where individuals so frequently learn only one lesson — not to stick their necks out.

It was in trying to teach what I had learned about treatment and counseling to graduate students at Ohio State University that I first began to realize that I had perhaps developed a distinctive point of view of my own, out of my experience. When I tried to crystallize some of these ideas, and present them in a paper at the University of Minnesota in December 1940, I found the reactions were very strong. It was my first experience of the fact that a new idea of mine, which to me can seem all shiny and glowing with potentiality, can to another person be a great threat. And to find myself the center of criticism, of arguments pro and con, was disconcerting and made me doubt and question. Nevertheless I felt I had something to contribute, and wrote the manuscript of *Counseling and Psychotherapy*, setting forth what I felt to be a somewhat more effective orientation to therapy.

Here again I realize with some amusement how little I have cared about being "realistic." When I submitted the manuscript, the publisher thought it was interesting and new, but wondered what classes would use it. I replied that I knew of only two — a course I was teaching and one in another university. The publisher felt I had made a grave mistake in not writing a text which would fit courses already being given. He was very dubious that he could sell 2,000 copies, which would be necessary to break even. It was only when I said I would take it to another publisher that he decided to make the gamble. I don't know which of us has been more surprised at its sales — 70,000 copies to date and still continuing.

RECENT YEARS

I believe that from this point to the present time my professional life — five years at Ohio State, twelve years at the University of Chicago, and four years at the University of Wisconsin — is quite well documented by what I have written. I will very briefly stress two or three points which have some significance for me.

I have learned to live in increasingly deep therapeutic relationships with an ever-widening range of clients. This can be and has been extremely rewarding. It can be and has been at times very frightening, when a deeply disturbed person seems to demand that I must be more than I am, in order to meet his need. Certainly the carrying on of therapy is something which demands continuing personal growth on the part of the therapist, and this is sometimes painful, even though in the long run rewarding.

I would also mention the steadily increasing importance which research has come to have for me. Therapy is the experience in which I can let myself go subjectively. Research is the experience in which I can stand off and try to view this rich subjective experience with objectivity, applying all the elegant methods of science to determine whether I have been deceiving myself. The conviction grows in me that we shall discover laws of personality and behavior which are as significant for human progress or human understanding as the law of gravity or the laws of thermodynamics.

In the last two decades I have become somewhat more accustomed to being fought over, but the reactions to my ideas continue to surprise me. From my point of view I have felt that I have always put forth my thoughts in a tentative manner, to be accepted or rejected by the reader or the student. But at different times and places psychologists, counselors, and educators have been moved to great wrath, scorn and criticism by my views. As this furore has tended to die down in these fields it has in recent years been renewed among psychiatrists, some of whom sense, in my way of working, a deep threat to many of their most cherished and unquestioned principles. And perhaps the storms of criticism are more than matched by the damage done by uncritical and unquestioning "disciples" — individuals who have acquired something of a new point of view for themselves and have gone forth to do battle with all and sundry, using as weapons both inaccurate and accurate understandings of me and my work. I have found it difficult to know, at times, whether I have been hurt more by my "friends" or my enemies.

Perhaps partly because of the troubling business of being struggled over, I have come to value highly the privilege of getting away, of being alone. It has seemed to me that my most fruitful periods of work are the times when I have been able to get completely away from what others think, from professional expectations and daily demands, and gain perspective on what I am doing. My wife and I have found isolated hideaways in Mexico and in the Caribbean where no one knows I am a psychologist; where painting, swimming, snorkeling, and capturing some of the scenery in color photography are my major activities. Yet in these spots, where no more than two to four hours a day goes for professional work, I have made most of whatever advances I have made in the last few years. I prize the privilege of being alone.

Some Significant Learnings

There, in very brief outline, are some of the externals of my professional life. But I would like to take you inside, to tell you some of the things I have learned from the thousands of hours I have spent working intimately with individuals in personal distress.

I would like to make it very plain that these are learnings which have significance for *me.* I do not know whether they would hold true for you. I have no desire to present them as a guide for anyone else. Yet I have found that when another person has been willing to tell me something of his inner directions this has been of value to me, if only in sharpening my realization that my directions are different. So it is in that spirit that I offer the learnings which follow. In each case I believe they became a part of my actions and inner convictions long before I realized them consciously. They are certainly scattered learnings, and incomplete. I can only say that they are and have been very important to me. I continually learn and relearn them. I frequently fail to act in terms of them, but later I wish that I had. Frequently I fail to see a new situation as one in which some of these learnings might apply.

They are not fixed. They keep changing. Some seem to be acquiring a stronger emphasis, others are perhaps less important to me than at one time, but they are all, to me, significant.

I will introduce each learning with a phrase or sentence which gives something of its personal meaning. Then I will elaborate on it a bit. There is not much organization to what follows except that the first learnings have to do mostly with relationships to others. There follow some that fall in the realm of personal values and convictions.

I might start off these several statements of significant learnings with a negative item. *In my relationships with persons I have found that it does not help, in the long run, to act as though I were something that I am not.* It does not help to act calm and pleasant when actually I am angry and critical. It does not help to act as though I know the answers when I do not. It does not help to act as though I were a loving person if actually, at the moment, I am hostile. It does not help for me to act as though I were full of assurance, if actually I am frightened and unsure. Even on a very simple level I have found that this statement seems to hold. It does not help for me to act as though I were well when I feel ill.

What I am saying here, put in another way, is that I have not found it to be helpful or effective in my relationships with other people to try to maintain a façade; to act in one way on the surface when I am experiencing something quite different underneath. It does not, I believe, make me helpful in my attempts to build up constructive relationships with other individuals. I would want to make it clear that while I feel I have learned this to be true, I have by no means adequately profited from it. In fact, it seems to me that most of the mistakes I make in personal relationships, most of the times in which I fail to be of help to other individuals, can be accounted for in terms of the fact that I have, for some defensive reason, behaved in one way at a surface level, while in reality my feelings run in a contrary direction.

A second learning might be stated as follows — *I find I am more effective when I can listen acceptantly to myself, and can be myself.* I feel that over the years I have learned to become more adequate in listening to *myself;* so that I know, somewhat more adequately than I used to, what I am feeling at any given moment — to be able to realize I *am* angry, or that I *do* feel rejecting toward this person; or that I feel very full of warmth and affection for this individual; or that I am bored and uninterested in what is going on; or that I am eager to understand this individual or that I am anxious and fearful in my relationship to this person. All of these diverse attitudes are feelings which I think I can listen to in myself. One way of putting this is that I feel I have become more adequate in letting myself *be* what I *am.* It becomes easier for me to accept myself as a decidedly imperfect person, who by no means functions at all times in the way in which I would like to function.

This must seem to some like a very strange direction in which to move. It seems to me to have value because the curious paradox is that when I accept myself as I am, then I change. I believe that I have learned this from my clients as well as within my own experience — that we cannot change, we cannot move away from what we are, until we thoroughly *accept* what we are. Then change seems to come about almost unnoticed.

Another result which seems to grow out of being myself is that relationships then become real. Real relationships have an exciting way of being vital and meaningful. If I can accept the fact that I am annoyed at or bored by this client or this student, then I am also much more likely to be able to accept his feelings in response. I can also accept the changed experience and the changed feelings which are then likely to occur in me and in him. Real relationships tend to change rather than to remain static.

So I find it effective to let myself be what I am in my attitudes; to know when I have reached my limit of endurance or of tolerance, and to accept that as a fact; to know when I desire to mold or manipulate people, and to accept that as a fact in myself. I would like to be as acceptant of these feelings as of feelings of warmth, interest, permissiveness, kindness, understanding, which are also a very real part of me. It is when I do accept all these attitudes as a fact, as a part of me, that my relationship with the other person then becomes what it is, and is able to grow and change most readily.

I come now to a central learning which has had a great deal of significance for me. I can state this learning as follows: *I have found it of enormous value when I can permit myself to understand another person.* The way in which I have worded this statement may seem strange to you. Is it necessary to *permit* oneself to understand another? I think that it is. Our first reaction to most of the statements which we hear from other people is an immediate evaluation, or judgment, rather than an understanding of it. When someone expresses some feeling or attitude or belief, our tendency is, almost immediately, to feel "That's right"; or "That's stupid"; "That's abnormal"; "That's unreasonable"; "That's incorrect"; "That's not nice." Very rarely do we permit ourselves to *understand* precisely what the meaning of his statement is to him. I believe this is because understanding is risky. If I let myself really understand another person, I might be changed by that understanding. And we all fear change. So as I say, it is not an easy thing to permit oneself to understand an individual, to enter thoroughly and completely and empathically into his frame of reference. It is also a rare thing.

To understand is enriching in a double way. I find when I am working with clients in distress, that to understand the bizarre world of a psychotic individual, or to understand and sense the attitudes of a person who feels that life is too tragic to bear, or to understand a man who feels that he is a worthless and inferior individual—each of these understandings somehow enriches me. I learn from these experiences in ways that change me, that make me a different and, I think, a more responsive person. Even more important perhaps, is the fact that my understanding of these individuals permits them to change. It permits them to accept their own fears and bizarre thoughts and tragic feelings and discouragements, as well as their moments of courage and kindness and love and sensitivity. And it is their experience as well as mine that when someone fully understands those feelings, this enables them to accept those feelings in themselves. Then they find both the feelings and themselves changing. Whether it is understanding a woman who feels that very literally she has a hook in her head by which others lead her about, or understanding a man who feels that no one is as lonely, no one is as separated from others as he, I find these understandings to be of value to me. But also, and even more importantly, to be understood has a very positive value to these individuals.

Here is another learning which has had importance for me. *I have found it enriching to open channels whereby others can communicate their feelings, their private perceptual worlds, to me.* Because understanding is rewarding, I would like to reduce the barriers between others and me, so that they can, if they wish, reveal themselves more fully.

In the therapeutic relationship there are a number of ways by which I can make it easier for the client to communicate himself. I can by my own attitudes create a safety in the relationship which makes such communication more possible. A sensitiveness of understanding which sees him as he is to himself, and accepts him as having those perceptions and feelings, helps too.

But as a teacher also I have found that I am enriched when I can open channels through which others can share themselves with me. So I try, often not too successfully, to create a climate in the classroom where feelings can be expressed, where people can differ—

with each other and with the instructor. I have also frequently asked for "reaction sheets" from students — in which they can express themselves individually and personally regarding the course. They can tell of the way it is or is not meeting their needs, they can express their feelings regarding the instructor, or can tell of the personal difficulties they are having in relation to the course. These reaction sheets have no relation whatsoever to their grade. Sometimes the same sessions of a course are experienced in diametrically opposite ways. One student says, "My feeling is one of indefinable revulsion with the tone of this class." Another, a foreign student, speaking of the same week of the same course says, "Our class follows the best, fruitful and scientific way of learning. But for people who have been taught for a long, long time, as we have, by the lecture type, authoritative method, this new procedure is ununderstandable. People like us are conditioned to hear the instructor, to keep passively our notes and memorize his reading assignments for the exams. There is no need to say that it takes long time for people to get rid of their habits regardless of whether or not their habits are sterile, infertile and barren." To open myself to these sharply different feelings has been a deeply rewarding thing.

I have found the same thing true in groups where I am the administrator, or perceived as the leader. I wish to reduce the need for fear or defensiveness, so that people can communicate their feelings freely. This has been most exciting, and has led me to a whole new view of what administration can be. But I cannot expand on that here.

There is another very important learning which has come to me in my counseling work. I can voice this learning very briefly. *I have found it highly rewarding when I can accept another person.*

I have found that truly to accept another person and his feelings is by no means an easy thing, any more than is understanding. Can I really permit another person to feel hostile toward me? Can I accept his anger as a real and legitimate part of himself? Can I accept him when he views life and its problems in a way quite different from mine? Can I accept him when he feels very positively toward me, admiring me and wanting to model himself after me? All this is involved in acceptance, and it does not come easy. I believe that it is an increasingly common pattern in our culture for each one of us to believe, "Every other person must feel and think and believe the same as I do." We find it very hard to permit our children or our parents or our spouses to feel differently than we do about particular issues or problems. We cannot permit our clients or our students to differ from us or to utilize their experience in their own individual ways. On a national scale, we cannot permit another nation to think or feel differently than we do. Yet it has come to seem to me that this separateness of individuals, the right of each individual to utilize his experience in his own way and to discover his own meanings in it, — this is one of the most priceless potentialities of life. Each person is an island unto himself, in a very real sense; and he can only build bridges to other islands if he is first of all willing to be himself and permitted to be himself. So I find that when I can accept another person, which means specifically accepting the feelings and attitudes and beliefs that he has as a real and vital part of him, then I am assisting him to become a person: and there seems to me great value in this.

The next learning I want to state may be difficult to communicate. It is this. *The more I am open to the realities in me and in the other person, the less do I find myself wishing to rush in to "fix things."* As I try to listen to myself and the experiencing going on in me, and the more I try to extend that same listening attitude to another person, the more respect I feel for the complex processes of life. So I become less and less inclined to hurry in to fix things, to set goals, to mold people, to manipulate and push them in the way that I would like them to go. I am much more content simply to be myself and to let another person be himself. I know very well that this must seem like a strange, almost an Oriental point of view. What is life for if we are not going to do things to people? What is life for if we are not going to mold them to our purposes? What is life for if we are not going to teach them the things that *we* think they should learn? What is life for if we are not going to make them think and feel as we do? How can anyone hold such an inactive point of view as the one I am expressing? I am sure that attitudes such as these must be a part of the reaction of many of you.

Yet the paradoxical aspect of my experience is that the more I am simply willing to be myself, in all this complexity of life and the more I am willing to understand and accept the realities in myself and in the other person, the more change seems to be stirred up. It is a very paradoxical thing — that to the degree that each one of us is willing to be himself, then he finds not only himself changing; but he finds that other people to whom he relates are also changing. At least this is a very vivid part of my experience, and one of the deepest things I think I have learned in my personal and professional life.

Let me turn now to some other learnings which are less concerned with relationships, and have more to do with my own actions and values. The first of these is very brief. *I can trust my experience.*

One of the basic things which I was a long time in realizing, and which I am still learning, is that when an activity *feels* as though it is valuable or worth doing, it *is* worth doing. Put another way, I have learned that my total organismic sensing of a situation is more trustworthy than my intellect.

All of my professional life I have been going in directions which others thought were foolish, and about which I have had many doubts myself. But I have never regretted moving in directions which "felt right," even though I have often felt lonely or foolish at the time.

I have found that when I have trusted some inner non-intellectual sensing, I have discovered wisdom in the move. In fact I have found that when I have followed one of these unconventional paths because it felt right or true, then in five or ten years many of my colleagues have joined me, and I no longer need to feel alone in it.

As I gradually come to trust my total reactions more deeply, I find that I can use them to guide my thinking. I have come to have more respect for those vague thoughts which occur in me from time to time, which *feel* as though they were significant. I am inclined to think that these unclear thoughts or hunches will lead me to important areas. I think of it as trusting the totality of my experience, which I have learned to suspect is wiser than my intellect. It is fallible I am sure, but I believe it to be less fallible than my conscious mind alone. My attitude is very well expressed by Max Weber, the artist, when he says. "In carrying on my own humble creative effort, I depend greatly upon that which I do not yet know, and upon that which I have not yet done."

Very closely related to this learning is a corollary that, *evaluation by others is not a guide for me.* The judgments of others, while they are to be listened to, and taken into account for what they are, can never be a guide for me. This has been a hard thing to learn. I remember how shaken I was, in the early days, when a scholarly thoughtful man who seemed to me a much more competent and knowledgeable psychologist than I, told me what a mistake I was making by getting interested in psychotherapy. It could never lead anywhere, and as a psychologist I would not even have the opportunity to practice it.

In later years it has sometimes jolted me a bit to learn that I am, in the eyes of some others, a fraud, a person practicing medicine without a license, the author of a very superficial and damaging sort of therapy, a power seeker, a mystic, etc. And I have been equally disturbed by equally extreme praise. But I have not been too much concerned because I have come to feel that only one person (at least in my lifetime, and perhaps ever) can know whether what I am doing is honest, thorough, open, and sound, or false and defensive and unsound, and I am that person. I am happy to get all sorts of evidence regarding what I am doing and criticism (both friendly and hostile) and praise (both sincere and fawning) are a part of such evidence. But to weigh this evidence and to determine its meaning and usefulness is a task I cannot relinquish to anyone else.

In view of what I have been saying the next learning will probably not surprise you. *Experience is, for me, the highest authority.* The touchstone of validity is my own experience. No other person's ideas, and none of my own ideas, are as authoritative as my experience. It is to experience that I must return again and again, to discover a closer approximation to truth as it is in the process of becoming in me.

Neither the Bible nor the prophets — neither Freud nor research — neither the revelations of God nor man — can take precedence over my own direct experience.

My experience is the more authoritative as it becomes more primary, to use the semanticist's term. Thus the hierarchy of experience would be most authoritative at its lowest level. If I read a theory of psychotherapy, and if I formulate a theory of psychotherapy based on my work with clients, and if I also have a direct experience of psychotherapy with a client, then the degree of authority increases in the order in which I have listed these experiences.

My experience is not authoritative because it is infallible. It is the basis of authority because it can always be checked in new primary ways. In this way its frequent error or fallibility is always open to correction.

Now another personal learning. *I enjoy the discovering of order in experience.* It seems inevitable that I seek for the meaning or the orderliness or lawfulness in any large body of experience. It is this kind of curiosity, which I find it very satisfying to pursue, which has led me to each of the major formulations I have made. It led me to search for the orderliness in all the conglomeration of things clinicians did for children, and out of that came my book on *The Clinical Treatment of the Problem Child.* It led me to formulate the general principles which seemed to be operative in psychotherapy, and that has led to several books and many articles. It has led me into research to test the various types of lawfulness which I feel I have encountered in my experience. It has enticed me to construct theories to bring together the orderliness of that which has already been experienced and to project this order forward into new and unexplored realms where it may be further tested.

Thus I have come to see both scientific research and the process of theory construction as being aimed toward the inward ordering of significant experience. Research is the persistent disciplined effort to make sense and order out of the phenomena of subjective experience. It is justified because it is satisfying to perceive the world as having order, and because rewarding results often ensue when one understands the orderly relationships which appear in nature.

So I have come to recognize that the reason I devote myself to research, and to the building of theory, is to satisfy a need for perceiving order and meaning, a subjective need which exists in me. I have, at times, carried on research for other reasons — to satisfy others, to convince opponents and sceptics, to get ahead professionally, to gain prestige, and for other unsavory reasons. These errors in judgment and activity have only served to convince me more deeply that there is only one sound reason for pursuing scientific activities, and that is to satisfy a need for meaning which is in me.

Another learning which cost me much to recognize, can be stated in four words. *The facts are friendly.*

It has interested me a great deal that most psychotherapists, especially the psychoanalysts, have steadily refused to make any scientific investigation of their therapy, or to permit others to do this. I can understand this reaction because I have felt it. Especially in our early investigations I can well remember the anxiety of waiting to see how the findings came out. Suppose our hypotheses were *disproved!* Suppose we were mistaken in our views! Suppose our opinions were not justified! At such times, as I look back, it seems to me that I regarded the facts as potential enemies, as possible bearers of disaster. I have perhaps been slow in coming to realize that the facts are *always* friendly. Every bit of evidence one can acquire, in any area, leads one that much closer to what is true. And being closer to the truth can never be a harmful or dangerous or unsatisfying thing. So while I still hate to readjust my thinking, still hate to give up old ways of perceiving and conceptualizing, yet at some deeper level I have, to a considerable degree, come to realize that these painful reorganizations are what is known as *learning*, and that though painful they always lead to a more satisfying because somewhat more accurate way of seeing life. Thus at the present time one of the most enticing areas for thought and speculation is an area where several of my pet ideas have *not* been upheld by the evidence. I feel if I can only puzzle my way through this problem that I will find a much more satisfying approximation to the truth. I feel sure the facts will be my friends.

Somewhere here I want to bring in a learning which has been most rewarding, because it makes me feel so deeply akin to others. I can word it this way. *What is most personal is most general.* There have been times when in talking with students or staff, or in my writing, I have expressed myself in ways so personal that I have felt I was expressing an attitude which it was probable no one else could understand, because it was so uniquely my own. Two written examples of this are the Preface to *Client-Centered Therapy* (regarded as most unsuitable by the publishers), and an article on "Persons or Science." In these instances I have almost invariably found that the very feeling which has seemed to me most private, most personal, and hence most incomprehensible by others, has turned out to be an expression for which there is a resonance in many other people. It has led me to believe that what is most personal and unique in each one of us is probably the very element which would, if it were shared or expressed, speak most deeply to others. This has helped me to understand artists and poets as people who have dared to express the unique in themselves.

There is one deep learning which is perhaps basic to all of the things I have said thus far. It has been forced upon me by more than twenty-five years of trying to be helpful to individuals in personal distress. It is simply this. *It has been my experience that persons have a basically positive direction.* In my deepest contacts with individuals in therapy, even those whose troubles are most disturbing, whose behavior has been most anti-social, whose feelings seem most abnormal, I find this to be true. When I can sensitively understand the feelings which they are expressing, when I am able to accept them as separate persons in their own right, then I find that they tend to move in certain directions. And what are these directions in which they tend to move? The words which I believe are most truly descriptive are words such as positive, constructive, moving toward self-actualization, growing toward maturity, growing toward socialization. I have come to feel that the more fully the individual is understood and accepted, the more he tends to drop the false fronts with which he has been meeting life, and the more he tends to move in a direction which is forward.

I would not want to be misunderstood on this. I do not have a Pollyanna view of human nature. I am quite aware that out of defensiveness and inner fear individuals can and do behave in ways which are incredibly cruel, horribly destructive, immature, regressive, anti-social, hurtful. Yet one of the most refreshing and invigorating parts of my experience is to work with such individuals and to discover the strongly positive directional tendencies which exist in them, as in all of us, at the deepest levels.

Let me bring this long list to a close with one final learning which can be stated very briefly. *Life, at its best, is a flowing, changing process in which nothing is fixed.* In my clients and in myself I find that when life is richest and most rewarding it is a flowing process. To experience this is both fascinating and a little frightening. I find I am at my best when I can let the flow of my experience carry me, in a direction which appears to be forward, toward goals of which I am but dimly aware. In thus floating with the complex stream of my experiencing, and in trying to understand its ever-changing complexity, it should be evident that there are no fixed points. When I am thus able to be in process, it is clear that there can be no closed system of beliefs, no unchanging set of principles which I hold. Life is guided by a changing understanding of and interpretation of my experience. It is always in process of becoming.

I trust it is clear now why there is no philosophy or belief or set of principles which I could encourage or persuade others to have or hold. I can only try to live by *my* interpretation of the current meaning of *my* experience, and try to give others the permission and freedom to develop their own inward freedom and thus their own meaningful interpretation of their own experience.

If there is such a thing as truth, this free individual process of search should, I believe, converge toward it. And in a limited way, this is also what I seem to have experienced.

PART II

How Can I Be of Help?

I have found a way of working with individuals which seems to have much constructive potential.

2

Some Hypotheses Regarding the Facilitation of Personal Growth

The three chapters which constitute Part II span a period of six years, from 1954 to 1960. Curiously, they span a large segment of the country in their points of delivery — Oberlin, Ohio; St. Louis, Missouri; and Pasadena, California. They also cover a period in which much research was accumulating, so that statements made tentatively in the first paper are rather solidly confirmed by the time of the third.

In the following talk given at Oberlin College in 1954 I was trying to compress into the briefest possible time the fundamental principles of psychotherapy which had been expressed at greater length in my books, (Counseling and Psychotherapy) *(1942) and* (Client-Centered Therapy) *(1951). It is of interest to me that I present the facilitating relationship, and the outcomes, with no description of, or even comment on, the process by which change comes about.*

TO BE FACED by a troubled, conflicted person who is seeking and expecting help, has always constituted a great challenge to me. Do I have the knowledge, the resources, the psychological strength, the skill — do I have whatever it takes to be of help to such an individual?

For more than twenty-five years I have been trying to meet this kind of challenge. It has caused me to draw upon every element of my professional background: the rigorous methods of personality measurement which I first learned at Teachers' College, Columbia; the Freudian psychoanalytic insights and methods of the Institute for Child Guidance where I worked as interne; the continuing developments in the field of clinical psychology, with which I have been closely associated; the briefer exposure to the work of Otto Rank, to the methods of psychiatric social work, and other resources too numerous to mention. But most of all it has meant a continual learning from my own experience and that of my colleagues at the Counseling Center as we have endeavored to discover for ourselves effective means of working with people in distress. Gradually I have developed a way of working which grows out of that experience, and which can be tested, refined, and reshaped by further experience and by research.

A General Hypothesis

One brief way of describing the change which has taken place in me is to say that in my early professional years I was asking the question, How can I treat, or cure, or change this person? Now I would phrase the question in this way: How can I provide a relationship which this person may use for his own personal growth?

It is as I have come to put the question in this second way that I realize that whatever I have learned is applicable to all of my human relationships, not just to working with clients with problems. It is for this reason that I feel it is possible that the learnings which have had meaning for me in my experience may have some meaning for you in your experience, since all of us are involved in human relationships.

Perhaps I should start with a negative learning. It has gradually been driven home to me that I cannot be of help to this troubled person by means of any intellectual or training procedure. No approach which relies upon knowledge, upon training, upon the acceptance of something that is *taught,* is of any use. These approaches seem so tempting and direct that I have, in the past, tried a great many of them. It is possible to explain a person to himself, to prescribe steps which should lead him forward, to train him in knowledge about a more satisfying mode of life. But such methods are, in my experience, futile and inconsequential. The most they can accomplish is some temporary change, which soon disappears, leaving the individual more than ever convinced of his inadequacy.

The failure of any such approach through the intellect has forced me to recognize that change appears to come about through experience in a relationship. So I am going to try to state very briefly and informally, some of the essential hypotheses regarding a helping relationship which have seemed to gain increasing confirmation both from experience and research.

I can state the overall hypothesis in one sentence, as follows. If I can provide a certain type of relationship, the other person will discover within himself the capacity to use that relationship for growth, and change and personal development will occur.

The Relationship

But what meaning do these terms have? Let me take separately the three major phrases in this sentence and indicate something of the meaning they have for me. What is this certain type of relationship I would like to provide?

I have found that the more that I can be genuine in the relationship, the more helpful it will be. This means that I need to be aware of my own feelings, in so far as possible, rather than presenting an outward façade of one attitude, while actually holding another attitude at a deeper or unconscious level. Being genuine also involves the willingness to be and to express, in my words and my behavior, the various feelings and attitudes which exist in me. It is only in this way that the relationship can have *reality*, and reality seems deeply important as a first condition. It is only by providing the genuine reality which is in me, that the other person can successfully seek for the reality in him. I have found this to be true even when the attitudes I feel are not attitudes with which I am pleased, or attitudes which seem conducive to a good relationship. It seems extremely important to be *real.*

As a second condition, I find that the more acceptance and liking I feel toward this individual, the more I will be creating a relationship which he can use. By acceptance I mean a warm regard for him as a person of unconditional self-worth — of value no matter what his condition, his behavior, or his feelings. It means a respect and liking for him as a separate person, a willingness for him to possess his own feelings in his own way. It means an acceptance of and regard for his attitudes of the moment, no matter how negative or positive, no matter how much they may contradict other attitudes he has held in the past. This acceptance of each fluctuating aspect of this other person makes it for him a relationship of warmth and safety, and the safety of being liked and prized as a person seems a highly important element in a helping relationship.

I also find that the relationship is significant to the extent that I feel a continuing desire to understand — a sensitive empathy with each of the client's feelings and communications as they seem to him at that moment. Acceptance does not mean much until it involves understanding. It is only as I *understand* the feelings and thoughts which seem so horrible to you, or so weak, or so sentimental, or so bizarre — it is only as I see them as you see them, and accept them and you, that you feel really free to explore all the hidden nooks and frightening crannies of your inner and often buried experience. This *freedom* is an important condition of the relationship. There is implied here a freedom to explore oneself at both conscious and unconscious levels, as rapidly as one can dare to embark on this dangerous quest. There is also a complete freedom from any type of moral or diagnostic evaluation, since all such evaluations are, I believe, always threatening.

Thus the relationship which I have found helpful is characterized by a sort of transparency on my part, in which my real feelings are evident; by an acceptance of this other person as a separate person with value in his own right; and by a deep empathic understanding which enables me to see his private world through his eyes. When these conditions are achieved, I become a companion to my client, accompanying him in the frightening search for himself, which he now feels free to undertake.

I am by no means always able to achieve this kind of relationship with another, and sometimes, even when I feel I have achieved it in myself, he may be too frightened to perceive what is being offered to him. But I would say that when I hold in myself the kind of attitudes I have described, and when the other person can to some degree experience these attitudes, then I believe that change and constructive personal development will *invariably* occur — and I include the word "invariably" only after long and careful consideration.

The Motivation for Change

So much for the relationship. The second phrase in my overall hypothesis was that the individual will discover within himself the capacity to use this relationship for growth. I will try to indicate something of the meaning which that phrase has for me. Gradually my experience has forced me to conclude that the individual has within himself the capacity and the tendency, latent if not evident, to move forward toward maturity. In a suitable psychological climate this tendency is released, and becomes actual rather than potential. It is evident in the capacity of the individual to understand those aspects of his life and of himself which are causing him pain and dissatisfaction, an understanding which probes beneath his conscious knowledge of himself into those experiences which he has hidden from himself because of their threatening nature. It shows itself in the tendency to reorganize his personality and his relationship to life in ways which are regarded as more mature. Whether one calls it a growth tendency, a drive toward self-actualization, or a forward-moving directional tendency, it is the mainspring of life, and is, in the last analysis, the tendency upon which all psychotherapy depends. It is the urge which is evident in all organic and human life — to expand, extend, become autonomous, develop, mature — the tendency to express and activate all the capacities of the organism, to the extent that such activation enhances the organism or the self. This tendency may become deeply buried under layer after layer of encrusted psychological defenses; it may be hidden behind elaborate façades which deny its existence; but it is my belief that it exists in every individual, and awaits only the proper conditions to be released and expressed.

The Outcomes

I have attempted to describe the relationship which is basic to constructive personality change. I have tried to put into words the type of capacity which the individual brings to such a relationship. The third phrase of my general statement was that change and personal development would occur. It is my hypothesis that in such a relationship the individual will reorganize himself at both the conscious and deeper levels of his personality in such a manner as to cope with life more constructively, more intelligently, and in a more socialized as well as a more satisfying way.

Here I can depart from speculation and bring in the steadily increasing body of solid research knowledge which is accumulating. We know now that individuals who live in such a relationship even for a relatively limited number of hours show profound and significant changes in personality, attitudes, and behavior, changes that do not occur in matched control groups. In such a relationship the individual becomes more integrated, more effective. He shows fewer of the characteristics which are usually termed neurotic or psychotic, and more of the characteristics of the healthy, well-functioning person. He changes his perception of himself, becoming more realistic in his views of self. He becomes more like the person he wishes to be. He values himself more highly. He is more self-confident and self-directing. He has a better understanding of himself, becomes more open to his experience, denies or represses less of his experience. He becomes more accepting in his attitudes toward others, seeing others as more similar to himself.

In his behavior he shows similar changes. He is less frustrated by stress, and recovers from stress more quickly. He becomes more mature in his everyday behavior as this is observed by friends. He is less defensive, more adaptive, more able to meet situations creatively.

These are some of the changes which we now know come about in individuals who have completed a series of counseling interviews in which the psychological atmosphere approximates the relationship I described. Each of the statements made is based upon objective evidence. Much more research needs to be done, but there can no longer be any doubt as to the effectiveness of such a relationship in producing personality change.

A Broad Hypothesis of Human Relationships

To me, the exciting thing about these research findings is not simply the fact that they give evidence of the efficacy of one form of psychotherapy, though that is by no means unimportant. The excitement comes from the fact that these findings justify an even broader hypothesis regarding all human relationships. There seems every reason to suppose that the therapeutic relationship is only one instance of interpersonal relations, and that the same lawfulness governs all such relationships. Thus it seems reasonable to hypothesize that if the parent creates with his child a psychological climate such as we have described, then the child will become more self-directing, socialized, and mature. To the extent that the teacher creates such a relationship with his class, the student will become a self-initiated learner, more original, more self-disciplined, less anxious and other-directed. If the administrator, or military or industrial leader, creates such a climate within his organization, then his staff will become more self-responsible, more creative, better able to adapt to new problems, more basically cooperative. It appears possible to me that we are seeing the emergence of a new field of human relationships, in which we may specify that if certain attitudinal conditions exist, then certain definable changes will occur.

Conclusion

Let me conclude by returning to a personal statement. I have tried to share with you something of what I have learned in trying to be of help to troubled, unhappy, maladjusted individuals. I have formulated the hypothesis which has gradually come to have meaning for me — not only in my relationship to clients in distress, but in all my human relationships. I have indicated that such research knowledge as we have supports this hypothesis, but that there is much more investigation needed. I should like now to pull together into one statement the conditions of this general hypothesis, and the effects which are specified.

If I can create a relationship characterized on my part:

by a genuineness and transparency, in which I am my real feelings;

by a warm acceptance of and prizing of the other person as a separate individual;

by a sensitive ability to see his world and himself as he sees them;

Then the other individual in the relationship:

will experience and understand aspects of himself which previously he has repressed;

will find himself becoming better integrated, more able to function effectively;

will become more similar to the person he would like to be;

will be more self-directing and self-confident;

will become more of a person, more unique and more self-expressive;

will be more understanding, more acceptant of others;

will be able to cope with the problems of life more adequately and more comfortably.

I believe that this statement holds whether I am speaking of my relationship with a client, with a group of students or staff members, with my family or children. It seems to me that we have here a general hypothesis which offers exciting possibilities for the development of creative, adaptive, autonomous persons.

3

The Characteristics of a Helping Relationship

I have long had the strong conviction — some might say it was an obsession — that the therapeutic relationship is only a special instance of interpersonal relationships in general, and that the same lawfulness governs all such relationships. This was the theme I chose to work out for myself when I was asked to give an address to the convention of the American Personnel and Guidance Association at St. Louis, in 1958.

Evident in this paper is the dichotomy between the objective and the subjective which has been such an important part of my experience during recent years. I find it very difficult to give a paper which is either wholly objective or wholly subjective. I like to bring the two worlds into close juxtaposition, even if I cannot fully reconcile them.

MY INTEREST IN PSYCHOTHERAPY has brought about in me an interest in every kind of helping relationship. By this term I mean a relationship in which at least one of the parties has the intent of promoting the growth, development, maturity, improved functioning, improved coping with life of the other. The other, in this sense, may be one individual or a group. To put it in another way, a helping relationship might be defined as one in which one of the participants intends that there should come about, in one or both parties, more appreciation of, more expression of, more functional use of the latent inner resources of the individual.

Now it is obvious that such a definition covers a wide range of relationships which usually are intended to facilitate growth. It would certainly include the relationship between mother and child, father and child. It would include the relationship between the physician and his patient. The relationship between teacher and pupil would often come under this definition, though some teachers would not have the promotion of growth as their intent. It includes almost all counselor-client relationships, whether we are speaking of educational counseling, vocational counseling, or personal counseling. In this last-mentioned area it would include the wide range of relationships between the psychotherapist and the hospitalized psychotic, the therapist and the troubled or neurotic individual, and the relationship between the therapist and the increasing number of so-called "normal" individuals who enter therapy to improve their own functioning or accelerate their personal growth.

These are largely one-to-one relationships. But we should also think of the large number of individual-group interactions which are intended as helping relationships. Some administrators intend that their relationship to their staff groups shall be of the sort which promotes growth, though other administrators would not have this purpose. The interaction between the group therapy leader and his group belongs here. So does the relationship of the community consultant to a community group. Increasingly the interaction between the industrial consultant and a management group is intended as a helping relationship. Perhaps this listing will point up the fact that a great many of the relationships in which we and others are involved fall within this category of interactions in which there is the purpose of promoting development and more mature and adequate functioning.

The Question

But what are the characteristics of those relationships which *do* help, which do facilitate growth? And at the other end of the scale is it possible to discern those characteristics which make a relationship unhelpful, even though it was the sincere intent to promote growth and development? It is to these questions, particularly the first, that I would like to take you with me over some of the paths I have explored, and to tell you where I am, as of now, in my thinking on these issues.

The Answers Given by Research

It is natural to ask first of all whether there is any empirical research which would give us an objective answer to these questions. There has not been a large amount of research in this area as yet, but what there is is stimulating and suggestive. I cannot report all of it but I would like to make a somewhat extensive sampling of the studies which have been done and state very briefly some of the findings. In so doing, oversimplification is necessary, and I am quite aware that I am not doing full justice to the researches I am mentioning, but it may give you the feeling that factual advances are being made and pique your curiosity enough to examine the studies themselves, if you have not already done so.

Studies of Attitudes

Most of the studies throw light on the attitudes on the part of the helping person which make a relationship growth-promoting or growth-inhibiting. Let us look at some of these.

A careful study of parent-child relationships made some years ago by Baldwin and others (1) at the Fels Institute contains interesting evidence. Of the various clusters of parental attitudes toward children, the "acceptant-democratic" seemed most growth-facilitating. Children of these parents with their warm and equalitarian attitudes showed an accelerated intellectual development (an increasing I.Q.), more originality, more emotional security and control, less excitability than children from other types of homes. Though somewhat slow initially in social development, they were, by the time they reached school age, popular, friendly, non-aggressive leaders.

Where parents' attitudes are classed as "actively rejectant" the children show a slightly decelerated intellectual development, relatively poor use of the abilities they do possess, and some lack of originality. They are emotionally unstable, rebellious, aggressive, and quarrelsome. The children of parents with other attitude syndromes tend in various respects to fall in between these extremes.

I am sure that these findings do not surprise us as related to child development. I would like to suggest that they probably apply to other relationships as well, and that the counselor or physician or administrator who is warmly emotional and expressive, respectful of the individuality of himself and of the other, and who exhibits a non-possessive caring, probably facilitates self-realization much as does a parent with these attitudes.

Let me turn to another careful study in a very different area. Whitehorn and Betz (2, 18) investigated the degree of success achieved by young resident physicians in working with schizophrenic patients on a psychiatric ward. They chose for special study the seven who had been outstandingly helpful, and seven whose patients had shown the least degree of improvement. Each group had treated about fifty patients. The investigators examined all the available evidence to discover in what ways the A group (the successful group) differed from the B group. Several significant differences were found. The physicians in the A group tended to see the schizophrenic in terms of the personal meaning which various behaviors had to the patient, rather than seeing him as a case history or a descriptive diagnosis. They also tended to work toward goals which were oriented to the personality of the patient, rather than such goals as reducing the symptoms or curing the disease. It was found that the helpful physicians, in their day by day interaction, primarily made use of active personal participation — a person-to-person relationship. They made less use of procedures which could be classed as "passive permissive." They were even less likely to use such procedures as interpretation, instruction or advice, or emphasis upon the practical care of the patient. Finally, they were much more likely than the B group to develop a relationship in which the patient felt trust and confidence in the physician.

Although the authors cautiously emphasize that these findings relate only to the treatment of schizophrenics, I am inclined to disagree. I suspect that similar facts would be found in a research study of almost any class of helping relationship.

Another interesting study focuses upon the way in which the person being helped perceives the relationship. Heine (11) studied individuals who had gone for psychotherapeutic help to psychoanalytic, client-centered, and Adlerian therapists. Regardless of the type of therapy, these clients report similar changes in themselves. But it is their perception of the relationship which is of particular interest to us here. When asked what accounted for the changes which had occurred, they expressed some differing explanations, depending on the orientation of the therapist. But their agreement on the major elements they had found helpful was even more significant. They indicated that these attitudinal elements in the relationship accounted for the changes which had taken place in themselves: the trust they had felt in the therapist; being understood by the therapist; the feeling of independence they had had in making choices and decisions. The therapist procedure which they had found most helpful was that the therapist clarified and openly stated feelings which the client had been approaching hazily and hesitantly.

There was also a high degree of agreement among these clients, regardless of the orientation of their therapists, as to what elements had been unhelpful in the relationship. Such therapist attitudes as lack of interest, remoteness or distance, and an over-degree of sympathy, were perceived as unhelpful. As to procedures, they had found it unhelpful when therapists had given direct specific advice regarding decisions or had emphasized past history rather than present problems. Guiding suggestions mildly given were perceived in an intermediate range — neither clearly helpful nor unhelpful.

Fiedler, in a much quoted study (7), found that expert therapists of differing orientations formed similar relationships with their clients. Less well known are the elements which characterized these relationships, differentiating them from the relationships formed by less expert therapists. These elements are: an ability to understand the client's meanings and feelings; a sensitivity to the client's attitudes; a warm interest without any emotional over-involvement.

A study by Quinn (14) throws light on what is involved in understanding the client's meanings and feelings. His study is surprising in that it shows that "understanding" of the client's meanings is essentially an attitude of *desiring* to understand. Quinn presented his judges only with recorded therapist statements taken from interviews. The raters had no knowledge of what the therapist was responding to or how the client reacted to his response. Yet it was found that the degree of understanding could be judged about as well from this material as from listening to the response in context. This seems rather conclusive evidence that it is an attitude of wanting to understand which is communicated.

As to the emotional quality of the relationship, Seeman (16) found that success in psychotherapy is closely associated with a strong and growing mutual liking and respect between client and therapist.

An interesting study by Dittes (4) indicates how delicate this relationship is. Using a physiological measure, the psychogalvanic reflex, to measure the anxious or threatened or alerted reactions of the client, Dittes correlated the deviations on this measure with judges' ratings of the degree of warm acceptance and permissiveness on the part of the therapist. It was found that whenever the therapist's attitudes changed even slightly in the direction of a lesser degree of acceptance, the number of abrupt GSR deviations significantly increased. Evidently when the relationship is experienced as less acceptant the organism organizes against threat, even at the physiological level.

Without trying fully to integrate the findings from these various studies, it can at least be noted that a few things stand out. One is the fact that it is the attitudes and feelings of the therapist, rather than his theoretical orientation, which is important. His procedures and techniques are less important than his attitudes. It is also worth noting that it is the way in which his attitudes and procedures are *perceived* which makes a difference to the client, and that it is this perception which is crucial.

"Manufactured" Relationships

Let me turn to research of a very different sort, some of which you may find rather abhorrent, but which nevertheless has a bearing upon the nature of a facilitating relationship. These studies have to do with what we might think of as manufactured relationships.

Verplanck (17), Greenspoon (8) and others have shown that operant conditioning of verbal behavior is possible in a relationship. Very briefly, if the experimenter says "Mhm," or "Good," or nods his head after certain types of words or statements, those classes of words tend to increase because of being reinforced. It has been shown that using such procedures one can bring about increases in such diverse verbal categories as plural nouns, hostile words, statements of opinion. The person is completely unaware that he is being influenced in any way by these reinforcers. The implication is that by such selective reinforcement we could bring it about that the other person in the relationship would be using whatever kinds of words and making whatever kinds of statements we had decided to reinforce.

Following still further the principles of operant conditioning as developed by Skinner and his group, Lindsley (12) has shown that a chronic schizophrenic can be placed in a "helping relationship" with a machine. The machine, somewhat like a vending machine, can be set to reward a variety of types of behaviors. Initially it simply rewards — with candy, a cigarette, or the display of a picture — the lever-pressing behavior of the patient. But it is possible to set it so that many pulls on the lever may supply a hungry kitten — visible in a separate enclosure — with a drop of milk. In this case the satisfaction is an altruistic one. Plans are being developed to reward similar social or altruistic behavior directed toward another patient, placed in the next room. The only limit to the kinds of behavior which might be rewarded lies in the degree of mechanical ingenuity of the experimenter.

Lindsley reports that in some patients there has been marked clinical improvement. Personally I cannot help but be impressed by the description of one patient who had gone from a deteriorated chronic state to being given free grounds privileges, this change being quite clearly associated with his interaction with the machine. Then the experimenter decided to study experimental extinction, which, put in more personal terms, means that no matter how many thousands of times the lever was pressed, no reward of any kind was forthcoming. The patient gradually regressed, grew untidy, uncommunicative, and his grounds privilege had to be revoked. This (to me) pathetic incident would seem to indicate that even in a relationship to a machine, trustworthiness is important if the relationship is to be helpful.

Still another interesting study of a manufactured relationship is being carried on by Harlow and his associates (10), this time with monkeys. Infant monkeys, removed from their mothers almost immediately after birth, are, in one phase of the experiment, presented with two objects. One might be termed the "hard mother," a sloping cylinder of wire netting with a nipple from which the baby may feed. The other is a "soft mother," a similar cylinder made of foam rubber and terry cloth. Even when an infant gets all his food from the "hard mother" he clearly and increasingly prefers the "soft mother." Motion pictures show that he definitely "relates" to this object, playing with it, enjoying it, finding security in clinging to it when strange objects are near, and using that security as a home base for venturing into the frightening world. Of the many interesting and challenging implications of this study, one seems reasonably clear. It is that no amount of direct food reward can take the place of certain perceived qualities which the infant appears to need and desire.

Two Recent Studies

Let me close this wide-ranging — and perhaps perplexing — sampling of research studies with an account of two very recent investigations. The first is an experiment conducted by Ends and Page (5). Working with hardened chronic hospitalized alcoholics who had been committed to a state hospital for sixty days, they tried three different methods of group psychotherapy. The method which they believed would be most effective was therapy based on a two-factor theory of learning; a client-centered approach was expected to be second; a psychoanalytically oriented approach was expected to be least efficient. Their results showed that the therapy based upon a learning theory approach was not only not helpful, but was somewhat deleterious. The outcomes were worse than those in the control group which had no therapy. The analytically oriented therapy produced some positive gain, and the client-centered group therapy was associated with the greatest amount of positive change. Follow-up data, extending over one and one-half years, confirmed the in-hospital findings, with the lasting improvement being greatest in the client-centered approach, next in the analytic, next the control group, and least in those handled by a learning theory approach.

As I have puzzled over this study, unusual in that the approach to which the authors were committed proved *least* effective, I find a clue, I believe, in the description of the therapy based on learning theory (13). Essentially it consisted (*a*) of pointing out and labelling the behaviors which had proved unsatisfying, (*b*) of exploring objectively with the client the reasons behind these behaviors, and (*c*) of establishing through re-education more effective problem-solving habits. But in all of this interaction the aim, as they formulated it, was to be impersonal. The therapist "permits as little of his own personality to intrude as is humanly possible." The "therapist stresses personal anonymity in his activities, i.e., he must studiously avoid impressing the patient with his own (therapist's) individual personality characteristics." To me this seems the most likely clue to the failure of this approach, as I try to interpret the facts in the light of the other research studies. To withhold one's self as a person and to deal with the other person as an object does not have a high probability of being helpful.

The final study I wish to report is one just being completed by Halkides (9). She started from a theoretical formulation of mine regarding the necessary and sufficient conditions for therapeutic change (15). She hypothesized that there would be a significant relationship between the extent of constructive personality change in the client and four counselor variables: (*a*) the degree of empathic understanding of the client manifested by the counselor; (*b*) the degree of positive affective attitude (unconditional positive regard) manifested by the counselor toward the client; (*c*) the extent to which the counselor is genuine, his words matching his own internal feeling; and (*d*) the extent to which the counselor's response matches the client's expression in the intensity of affective expression.

To investigate these hypotheses she first selected, by multiple objective criteria, a group of ten cases which could be classed as "most successful" and a group of ten "least successful" cases. She then took an early and late recorded interview from each of these cases. On a random basis she picked nine client-counselor interaction units — a client statement and a counselor response — from each of these interviews. She thus had nine early interactions and nine later interactions from each case. This gave her several hundred units which were now placed in random order. The units from an early interview of an unsuccessful case might be followed by the units from a late interview of a successful case, etc.

Three judges, who did not know the cases or their degree of success, or the source of any given unit, now listened to this material four different times. They rated each unit on a seven point scale, first as to the degree of empathy, second as to the counselor's positive attitude toward the client, third as to the counselor's congruence or genuineness, and fourth as to the degree to which the counselor's response matched the emotional intensity of the client's expression.

I think all of us who knew of the study regarded it as a very bold venture. Could judges listening to single units of interaction possibly make any reliable rating of such subtle qualities as I have mentioned? And even if suitable reliability could be obtained, could eighteen counselor-client interchanges from each case — a minute sampling of the hundreds or thousands of such interchanges which occurred in each case — possibly bear any relationship to the therapeutic outcome? The chance seemed slim.

The findings are surprising. It proved possible to achieve high reliability between the judges, most of the inter-judge correlations being in the 0.80's or 0.90's, except on the last variable. It was found that a high degree of empathic understanding was significantly associated, at a .001 level, with the more successful cases. A high degree of unconditional positive regard was likewise associated with the more successful cases, at the .001 level. Even the rating of the counselor's genuineness or congruence — the extent to which his words matched his feelings — was associated with the successful outcome of the case, and again at the .001 level of significance. Only in the investigation of the matching intensity of affective expression were the results equivocal.

It is of interest too that high ratings of these variables were not associated more significantly with units from later interviews than with units from early interviews. This means that the counselor's attitudes were quite constant throughout the interviews. If he was highly empathic, he tended to be so from first to last. If he was lacking in genuineness, this tended to be true of both early and late interviews.

As with any study, this investigation has its limitations. It is concerned with a certain type of helping relationship, psychotherapy. It investigated only four variables thought to be significant. Perhaps there are many others. Nevertheless it represents a significant advance in the study of helping relationships. Let me try to state the findings in the simplest possible fashion. It seems to indicate that the quality of the counselor's interaction with a client can be satisfactorily judged on the basis of a very small sampling of his behavior. It also means that if the counselor is congruent or transparent, so that his words are in line with his feelings rather than the two being discrepant; if the counselor likes the client, unconditionally; and if the counselor understands the essential feelings of the client as they seem to the client — then there is a strong probability that this will be an effective helping relationship.

Some Comments

These then are some of the studies which throw at least a measure of light on the nature of the helping relationship. They have investigated different facets of the problem. They have approached it from very different theoretical contexts. They have used different methods. They are not directly comparable. Yet they seem to me to point to several statements which may be made with some assurance. It seems clear that relationships which are helpful have different characteristics from relationships which are unhelpful. These differential characteristics have to do primarily with the attitudes of the helping person on the one hand and with the perception of the relationship by the "helpee" on the other. It is equally clear that the studies thus far made do not give us any final answers as to what is a helping relationship, nor how it is to be formed.

How Can I Create a Helping Relationship?

I believe each of us working in the field of human relationships has a similar problem in knowing how to use such research knowledge. We cannot slavishly follow such findings in a mechanical way or we destroy the personal qualities which these very studies show to be valuable. It seems to me that we have to use these studies, testing them against our own experience and forming new and further personal hypotheses to use and test in our own further personal relationships.

So rather than try to tell you how you should use the findings I have presented I should like to tell you the kind of questions which these studies and my own clinical experience raise for me, and some of the tentative and changing hypotheses which guide my behavior as I enter into what I hope may be helping relationships, whether with students, staff, family, or clients. Let me list a number of these questions and considerations.

1. Can I *be* in some way which will be perceived by the other person as trustworthy, as dependable or consistent in some deep sense? Both research and experience indicate that this is very important, and over the years I have found what I believe are deeper and better ways of answering this question. I used to feel that if I fulfilled all the outer conditions of trustworthiness — keeping appointments, respecting the confidential nature of the interviews, etc. — and if I acted consistently the same during the interviews, then this condition would be fulfilled. But experience drove home the fact that to act consistently acceptant, for example, if in fact I was feeling annoyed or skeptical or some other non-acceptant feeling, was certain in the long run to be perceived as inconsistent or untrustworthy. I have come to recognize that being trustworthy does not demand that I be rigidly consistent but that I be dependably real. The term "congruent" is one I have used to describe the way

I would like to be. By this I mean that whatever feeling or attitude I am experiencing would be matched by my awareness of that attitude. When this is true, then I am a unified or integrated person in that moment, and hence I can *be* whatever I deeply *am.* This is a reality which I find others experience as dependable.

2. A very closely related question is this: Can I be expressive enough as a person that what I am will be communicated unambiguously? I believe that most of my failures to achieve a helping relationship can be traced to unsatisfactory answers to these two questions. When I am experiencing an attitude of annoyance toward another person but am unaware of it, then my communication contains contradictory messages. My words are giving one message, but I am also in subtle ways communicating the annoyance I feel and this confuses the other person and makes him distrustful, though he too may be unaware of what is causing the difficulty. When as a parent or a therapist or a teacher or an administrator I fail to listen to what is going on in me, fail because of my own defensiveness to sense my own feelings, then this kind of failure seems to result. It has made it seem to me that the most basic learning for anyone who hopes to establish any kind of helping relationship is that it is safe to be transparently real. If in a given relationship I am reasonably congruent, if no feelings relevant to the relationship are hidden either to me or the other person, then I can be almost sure that the relationship will be a helpful one.

One way of putting this which may seem strange to you is that if I can form a helping relationship to myself — if I can be sensitively aware of and acceptant toward my own feelings — then the likelihood is great that I can form a helping relationship toward another.

Now, acceptantly to be what I am, in this sense, and to permit this to show through to the other person, is the most difficult task I know and one I never fully achieve. But to realize that this *is* my task has been most rewarding because it has helped me to find what has gone wrong with interpersonal relationships which have become snarled and to put them on a constructive track again. It has meant that if I am to facilitate the personal growth of others in relation to me, then I must grow, and while this is often painful it is also enriching.

3. A third question is: Can I let myself experience positive attitudes toward this other person — attitudes of warmth, caring, liking, interest, respect? It is not easy. I find in myself, and feel that I often see in others, a certain amount of fear of these feelings. We are afraid that if we let ourselves freely experience these positive feelings toward another we may be trapped by them. They may lead to demands on us or we may be disappointed in our trust, and these outcomes we fear. So as a reaction we tend to build up distance between ourselves and others — aloofness, a "professional" attitude, an impersonal relationship.

I feel quite strongly that one of the important reasons for the professionalization of every field is that it helps to keep this distance. In the clinical areas we develop elaborate diagnostic formulations, seeing the person as an object. In teaching and in administration we develop all kinds of evaluative procedures, so that again the person is perceived as an object. In these ways, I believe, we can keep ourselves from experiencing the caring which would exist if we recognized the relationship as one between two persons. It is a real achievement when we can learn, even in certain relationships or at certain times in those relationships, that it is safe to care, that it is safe to relate to the other as a person for whom we have positive feelings.

4. Another question the importance of which I have learned in my own experience is: Can I be strong enough as a person to be separate from the other? Can I be a sturdy respecter of my own feelings, my own needs, as well as his? Can I own and, if need be, express my own feelings as something belonging to me and separate from his feelings? Am I strong enough in my own separateness that I will not be downcast by his depression, frightened by his fear, nor engulfed by his dependency? Is my inner self hardy enough to realize that I am not destroyed by his anger, taken over by his need for dependence, nor enslaved by his love, but that I exist separate from him with feelings and rights of my own? When I can freely feel this strength of being a separate person, then I find that I can let myself go much more deeply in understanding and accepting him because I am not fearful of losing myself.

5. The next question is closely related. Am I secure enough

within myself to permit him his separateness? Can I permit him to be what he is — honest or deceitful, infantile or adult, despairing or over-confident? Can I give him the freedom to be? Or do I feel that he should follow my advice, or remain somewhat dependent on me, or mold himself after me? In this connection I think of the interesting small study by Farson (6) which found that the less well adjusted and less competent counselor tends to induce conformity to himself, to have clients who model themselves after him. On the other hand, the better adjusted and more competent counselor can interact with a client through many interviews without interfering with the freedom of the client to develop a personality quite separate from that of his therapist. I should prefer to be in this latter class, whether as parent or supervisor or counselor.

6. Another question I ask myself is: Can I let myself enter fully into the world of his feelings and personal meanings and see these as he does? Can I step into his private world so completely that I lose all desire to evaluate or judge it? Can I enter it so sensitively that I can move about in it freely, without trampling on meanings which are precious to him? Can I sense it so accurately that I can catch not only the meanings of his experience which are obvious to him, but those meanings which are only implicit, which he sees only dimly or as confusion? Can I extend this understanding without limit? I think of the client who said, "Whenever I find someone who understands a *part* of me at the time, then it never fails that a point is reached where I know they're *not* understanding me again . . . What I've looked for so hard is for someone to understand."

For myself I find it easier to feel this kind of understanding, and to communicate it, to individual clients than to students in a class or staff members in a group in which I am involved. There is a strong temptation to set students "straight," or to point out to a staff member the errors in his thinking. Yet when I can permit myself to understand in these situations, it is mutually rewarding. And with clients in therapy, I am often impressed with the fact that even a minimal amount of empathic understanding — a bumbling and faulty attempt to catch the confused complexity of the client's meaning — is helpful, though there is no doubt that it is most helpful when I can see and formulate clearly the meanings in his experiencing which for him have been unclear and tangled.

7. Still another issue is whether I can be acceptant of each facet of this other person which he presents to me. Can I receive him as he is? Can I communicate this attitude? Or can I only receive him conditionally, acceptant of some aspects of his feelings and silently or openly disapproving of other aspects? It has been my experience that when my attitude is conditional, then he cannot change or grow in those respects in which I cannot fully receive him. And when — afterward and sometimes too late — I try to discover why I have been unable to accept him in every respect, I usually discover that it is because I have been frightened or threatened in myself by some aspect of his feelings. If I am to be more helpful, then I must myself grow and accept myself in these respects.

8. A very practical issue is raised by the question: Can I act with sufficient sensitivity in the relationship that my behavior will not be perceived as a threat? The work we are beginning to do in studying the physiological concomitants of psychotherapy confirms the research by Dittes in indicating how easily individuals are threatened at a physiological level. The psychogalvanic reflex — the measure of skin conductance — takes a sharp dip when the therapist responds with some word which is just a little stronger than the client's feelings. And to a phrase such as, "My you *do* look upset," the needle swings almost off the paper. My desire to avoid even such minor threats is not due to a hypersensitivity about my client. It is simply due to the conviction based on experience that if I can free him as completely as possible from external threat, then he can begin to experience and to deal with the internal feelings and conflicts which he finds threatening within himself.

9. A specific aspect of the preceding question but an important one is: Can I free him from the threat of external evaluation? In almost every phase of our lives — at home, at school, at work — we find ourselves under the rewards and punishments of external judgments. "That's good"; "that's naughty." "That's worth an A"; "that's a failure." "That's good counseling"; "that's poor counseling." Such judgments are a part of our lives from infancy to old age. I believe they have a certain social usefulness to institutions and organizations such as schools and professions. Like everyone else

I find myself all too often making such evaluations. But, in my experience, they do not make for personal growth and hence I do not believe that they are a part of a helping relationship. Curiously enough a positive evaluation is as threatening in the long run as a negative one, since to inform someone that he is good implies that you also have the right to tell him he is bad. So I have come to feel that the more I can keep a relationship free of judgment and evaluation, the more this will permit the other person to reach the point where he recognizes that the locus of evaluation, the center of responsibility, lies within himself. The meaning and value of his experience is in the last analysis something which is up to him, and no amount of external judgment can alter this. So I should like to work toward a relationship in which I am not, even in my own feelings, evaluating him. This I believe can set him free to be a self-responsible person.

10. One last question: Can I meet this other individual as a person who is in process of *becoming*, or will I be bound by his past and by my past? If, in my encounter with him, I am dealing with him as an immature child, an ignorant student, a neurotic personality, or a psychopath, each of these concepts of mine limits what he can be in the relationship. Martin Buber, the existentialist philosopher of the University of Jerusalem, has a phrase, "confirming the other," which has had meaning for me. He says "Confirming means . . . accepting the whole potentiality of the other. . . . I can recognize in him, know in him, the person he has been . . . *created* to become. . . . I confirm him in myself, and then in him, in relation to this potentiality that . . . can now be developed, can evolve" (3). If I accept the other person as something fixed, already diagnosed and classified, already shaped by his past, then I am doing my part to confirm this limited hypothesis. If I accept him as a process of becoming, then I am doing what I can to confirm or make real his potentialities.

It is at this point that I see Verplanck, Lindsley, and Skinner, working in operant conditioning, coming together with Buber, the philosopher or mystic. At least they come together in principle, in an odd way. If I see a relationship as only an opportunity to reinforce certain types of words or opinions in the other, then I tend to confirm him as an object — a basically mechanical, manipulable object. And if I see this as his potentiality, he tends to act in ways which support this hypothesis. If, on the other hand, I see a relationship as an opportunity to "reinforce" *all* that he is, the person that he is with all his existent potentialities, then he tends to act in ways which support *this* hypothesis. I have then — to use Buber's term — confirmed him as a living person, capable of creative inner development. Personally I prefer this second type of hypothesis.

Conclusion

In the early portion of this paper I reviewed some of the contributions which research is making to our knowledge *about* relationships. Endeavoring to keep that knowledge in mind I then took up the kind of questions which arise from an inner and subjective point of view as I enter, as a person, into relationships. If I could, in myself, answer all the questions I have raised in the affirmative, then I believe that any relationships in which I was involved would be helping relationships, would involve growth. But I cannot give a positive answer to most of these questions. I can only work in the direction of the positive answer.

This has raised in my mind the strong suspicion that the optimal helping relationship is the kind of relationship created by a person who is psychologically mature. Or to put it in another way, the degree to which I can create relationships which facilitate the growth of others as separate persons is a measure of the growth I have achieved in myself. In some respects this is a disturbing thought, but it is also a promising or challenging one. It would indicate that if I am interested in creating helping relationships I have a fascinating lifetime job ahead of me, stretching and developing my potentialities in the direction of growth.

I am left with the uncomfortable thought that what I have been working out for myself in this paper may have little relationship to your interests and your work. If so, I regret it. But I am at least partially comforted by the fact that all of us who are working in the field of human relationships and trying to understand the basic orderliness of that field are engaged in the most crucial enterprise in today's world. If we are thoughtfully trying to understand our tasks as administrators, teachers, educational counselors, vocational counselors, therapists, then we are working on the problem which will determine the future of this planet. For it is not upon the physical sciences that the future will depend. It is upon us who are trying to understand and deal with the interactions between human beings — who are trying to create helping relationships. So I hope that the questions I ask of myself will be of some use to you in gaining understanding and perspective as you endeavor, in your way, to facilitate growth in your relationships.

References

1. Baldwin, A. L., J. Kalhorn, and F. H. Breese. Patterns of parent behavior. *Psychol. Monogr.*, 1945, *58*, No. 268, 1–75.

2. Betz, B. J., and J. C. Whitehorn. The relationship of the therapist to the outcome of therapy in schizophrenia. *Psychiat. Research Reports #5. Research techniques in schizophrenia.* Washington, D.C., American Psychiatric Association, 1956, 89–117.

3. Buber, M., and C. Rogers. Transcription of dialogue held April 18, 1957, Ann Arbor, Mich. Unpublished manuscript.

4. Dittes, J. E. Galvanic skin response as a measure of patient's reaction to therapist's permissiveness. *J. Abnorm. & Soc. Psychol.*, 1957, *55*, 295–303.

5. Ends, E. J., and C. W. Page. A study of three types of group psychotherapy with hospitalized male inebriates. *Quar. J. Stud. Alcohol*, 1957. *18*, 263–277.

6. Farson, R. E. Introjection in the psychotherapeutic relationship. Unpublished doctoral dissertation, University of Chicago, 1955.

7. Fiedler, F. E. Quantitative studies on the role of therapists feelings toward their patients. In Mowrer, O. H. (Ed.), *Psychotherapy: theory and research.* New York: Ronald Press, 1953, Chap. 12.

8. Greenspoon, J. The reinforcing effect of two spoken sounds on the frequency of two responses. *Amer. J. Psychol.*, 1955, *68*, 409–416.

9. Halkides, G. An experimental study of four conditions necessary for therapeutic change. Unpublished doctoral dissertation, University of Chicago, 1958.

10. Harlow, H. F. The nature of love. *Amer. Psychol.*, 1958, *13*, 673–685.

11. Heine, R. W. A comparison of patients' reports on psychotherapeutic experience with psychoanalytic, nondirective, and Adlerian therapists. Unpublished doctoral dissertation, University of Chicago, 1950.

12. Lindsley, O. R. Operant conditioning methods applied to research in chronic schizophrenia. *Psychiat. Research Reports #5. Research techniques in schizophrenia.* Washington, D.C.: American Psychiatric Association, 1956, 118–153.

13. Page, C. W., and E. J. Ends. A review and synthesis of the literature suggesting a psychotherapeutic technique based on two-factor learning theory. Unpublished manuscript, loaned to the writer.

14. Quinn, R. D. Psychotherapists' expressions as an index to the quality of early therapeutic relationships. Unpublished doctoral dissertation, University of Chicago, 1950.

15. Rogers, C. R. The necessary and sufficient conditions of psychotherapeutic personality change. *J. Consult. Psychol.*, 1957, *21*, 95–103.

16. Seeman, J. Counselor judgments of therapeutic process and outcome. In Rogers, C. R., and R. F. Dymond, (Eds.). *Psychotherapy and personality change.* University of Chicago Press, 1954, Chap. 7.

17. Verplanck, W. S. The control of the content of conversation: reinforcement of statements of opinion. *J. Abnorm. & Soc. Psychol.*, 1955, *51*, 668–676.

18. Whitehorn, J. C., and B. J. Betz. A study of psychotherapeutic relationships between physicians and schizophrenic patients. *Amer. J. Psychiat.*, 1954, *111*, 321–331.

4

What We Know About Psychotherapy — Objectively and Subjectively

In the spring of 1960 I was invited to the California Institute of Technology as a visitor in their "Leaders of America" program, sponsored by the Cal Tech YMCA, which arranges most of the cultural programs for the Institute. As one part of this four-day visit I was asked to talk to a forum of faculty and staff. I was eager to speak of psychotherapy in a way which would make sense to physical scientists, and it seemed to me a summary of the research findings in regard to therapy might communicate. On the other hand I wished to make very clear that the personal subjective relationship is at least an equally fundamental part of therapeutic change. So I endeavored to present both sides. I have made some changes in the paper, but this is essentially what I presented to the audience at Cal Tech.

I was pleased that the presentation seemed well received, but I have been even more pleased that since that time a number of individuals who have experienced therapy have read the manuscript and seem highly enthusiastic about the description (in the second half of the paper) of the client's inner experience of therapy. This gratifies me, because I am especially eager to capture the way therapy feels and seems to the client.

In the field of psychotherapy considerable progress has been made in the last decade in measuring the outcomes of therapy in the personality and behavior of the client. In the last two or three years additional progress has been made in identifying the basic conditions in the therapeutic relationship which bring about therapy, which facilitate personal development in the direction of psychological maturity. Another way of saying this is that we have made progress in determining those ingredients in a relationship which promote personal growth.

Psychotherapy does not supply the motivation for such development or growth. This seems to be inherent in the organism, just as we find a similar tendency in the human animal to develop and mature physically, provided minimally satisfactory conditions are provided. But therapy does play an extremely important part in releasing and facilitating the tendency of the organism toward psychological development or maturity, when this tendency has been blocked.

Objective Knowledge

I would like, in the first part of this talk, to summarize what we know of the conditions which facilitate psychological growth, and something of what we know of the process and characteristics of that psychological growth. Let me explain what I mean when I say that I am going to summarize what we "know." I mean that I will limit my statements to those for which we have objective empirical evidence. For example, I will talk about the conditions of psychological growth. For each statement one or more studies could be cited in which it was found that changes occurred in the individual when these conditions were present which did not occur in situations where these conditions were absent, or were present to a much lesser degree. As one investigator states, we have made progress in identifying the primary change-producing agents which facilitate the alteration of personality and of behavior in the direction of personal development. It should of course be added that this knowledge, like all scientific knowledge, is tentative and surely incomplete, and is certain to be modified, contradicted in part, and supplemented by the painstaking work of the future. Nevertheless there is no reason to be apologetic for the small but hard-won knowledge which we currently possess.

I would like to give this knowledge which we have gained in the very briefest fashion, and in everyday language.

It has been found that personal change is facilitated when the psychotherapist is what he *is*, when in the relationship with his client he is genuine and without "front" or façade, openly being the feelings and attitudes which at that moment are flowing *in* him. We have coined the term "congruence" to try to describe this condition. By this we mean that the feelings the therapist is experiencing are available to him, available to his awareness, and he is able to live these feelings, be them, and able to communicate them if appropriate. No one fully achieves this condition, yet the more the therapist is able to listen acceptantly to what is going on within himself, and the more he is able to be the complexity of his feelings, without fear, the higher the degree of his congruence.

To give a commonplace example, each of us senses this quality in people in a variety of ways. One of the things which offends us about radio and TV commercials is that it is often perfectly evident from the tone of voice that the announcer is "putting on," playing a role, saying something he doesn't feel. This is an example of incongruence. On the other hand each of us knows individuals whom we somehow trust because we sense that they are being what they are, that we are dealing with the person himself, not with a polite or professional front. It is this quality of congruence which we sense which research has found to be associated with successful therapy.

The more genuine and congruent the therapist in the relationship, the more probability there is that change in personality in the client will occur.

Now the second condition. When the therapist is experiencing a warm, positive and acceptant attitude toward what *is* in the client, this facilitates change. It involves the therapist's genuine willingness for the client to be whatever feeling is going on in him at that moment, — fear, confusion, pain, pride, anger, hatred, love, courage, or awe. It means that the therapist cares for the client, in a non-possessive way. It means that he prizes the client in a total rather than a conditional way. By this I mean that he does not simply accept the client when he is behaving in certain ways, and disapprove of him when he behaves in other ways. It means an outgoing positive feeling without reservations, without evaluations. The term we have come to use for this is unconditional positive regard. Again research studies show that the more this attitude is experienced by the therapist, the more likelihood there is that therapy will be successful.

The third condition we may call empathic understanding. When the therapist is sensing the feelings and personal meanings which the client is experiencing in each moment, when he can perceive these from "inside," as they seem to the client, and when he can successfully communicate something of that understanding to his client, then this third condition is fulfilled.

I suspect each of us has discovered that this kind of understanding is extremely rare. We neither receive it nor offer it with any great frequency. Instead we offer another type of understanding which is very different. "I understand what is wrong with you"; "I understand what makes you act that way"; or "I too have experienced your trouble and I reacted very differently"; these are the types of understanding which we usually offer and receive, an evaluative understanding from the outside. But when someone understands how it feels and seems to be *me*, without wanting to analyze me or judge me, then I can blossom and grow in that climate. And research bears out this common observation. When the therapist can grasp the moment-to-moment experiencing which occurs in the inner world of the client as the client sees it and feels it, without losing the separateness of his own identity in this empathic process, then change is likely to occur.

Studies with a variety of clients show that when these three conditions occur in the therapist, and when they are to some degree perceived by the client, therapeutic movement ensues, the client finds himself painfully but definitely learning and growing, and both he and his therapist regard the outcome as successful. It seems from our studies that it is attitudes such as these rather than the therapist's technical knowledge and skill, which are primarily responsible for therapeutic change.

The Dynamics of Change

You may well ask, "But why does a person who is seeking help change for the better when he is involved, over a period of time, in a relationship with a therapist which contains these elements? How does this come about?" Let me try very briefly to answer this question.

The reactions of the client who experiences for a time the kind of therapeutic relationship which I have described are a reciprocal of the therapist's attitudes. In the first place, as he finds someone else listening acceptantly to his feelings, he little by little becomes able to listen to himself. He begins to receive the communications from within himself — to realize that he *is* angry, to recognize when he is frightened, even to realize when he is feeling courageous. As he becomes more open to what is going on within him he becomes able to listen to feelings which he has always denied and repressed. He can listen to feelings which have seemed to him so terrible, or so disorganizing, or so abnormal, or so shameful, that he has never been able to recognize their existence in himself.

While he is learning to listen to himself he also becomes more acceptant of himself. As he expresses more and more of the hidden and awful aspects of himself, he finds the therapist showing a consistent and unconditional positive regard for him and his feelings. Slowly he moves toward taking the same attitude toward himself, accepting himself as he is, and therefore ready to move forward in the process of becoming.

And finally as he listens more accurately to the feelings within, and becomes less evaluative and more acceptant toward himself, he also moves toward greater congruence. He finds it possible to move out from behind the façades he has used, to drop his defensive behaviors, and more openly to be what he truly is. As these changes occur, as he becomes more self-aware, more self-acceptant, less defensive and more open, he finds that he is at last free to change and grow in the directions natural to the human organism.

THE PROCESS

Now let me put something of this process in factual statements, each statement borne out by empirical research. We know that the client shows movement on each of a number of continua. Starting from wherever he may be on each continuum I will mention, he moves toward the upper end.

In regard to feelings and personal meanings, he moves away from a state in which feelings are unrecognized, unowned, unexpressed. He moves toward a flow in which ever-changing feelings are experienced in the moment, knowingly and acceptingly, and may be accurately expressed.

The process involves a change in the manner of his experiencing. Initially he is remote from his experiencing. An example would be the intellectualizing person who talks about himself and his feelings in abstractions, leaving you wondering what is *actually* going on within him. From such remoteness he moves toward an immediacy of experiencing in which he lives openly *in* his experiencing, and knows that he can turn to it to discover its current meanings.

The process involves a loosening of the cognitive maps of experience. From construing experience in rigid ways, which are perceived as external facts, the client moves toward developing changing, loosely held construings of meaning in experience, constructs which are modifiable by each new experience.

In general, the evidence shows that the process moves away from fixity, remoteness from feelings and experience, rigidity of self-concept, remoteness from people, impersonality of functioning. It moves toward fluidity, changingness, immediacy of feelings and experience, acceptance of feelings and experience, tentativeness of constructs, discovery of a changing self in one's changing experience, realness and closeness of relationships, a unity and integration of functioning.

We are continually learning more about this process by which change comes about, and I am not sure that this very brief summary conveys much of the richness of our findings.

The Results of Therapy

But let me turn to the outcomes of therapy, to the relatively lasting changes which occur. As in the other things I have said I will limit myself to statements borne out by research evidence. The client changes and reorganizes his concept of himself. He moves away from perceiving himself as unacceptable to himself, as unworthy of respect, as having to live by the standards of others. He moves toward a conception of himself as a person of worth, as a self-directing person, able to form his standards and values upon the basis of his own experience. He develops much more positive attitudes toward himself. One study showed that at the beginning of therapy current attitudes toward self were four to one negative, but in the final fifth of therapy self-attitudes were twice as often positive as negative. He becomes less defensive, and hence more open to his experience of himself and of others. He becomes more realistic and differentiated in his perceptions. He improves in his psychological adjustment, whether this is measured by the Rorschach test, the Thematic Apperception Test, the counselor's rating, or other indices. His aims and ideals for himself change so that they are more achievable. The initial discrepancy between the self that he is and the self that he wants to be is greatly diminished. Tension of all types is reduced — physiological tension, psychological discomfort, anxiety. He perceives other individuals with more realism and more acceptance. He describes his own behavior as being more mature and, what is more important, he is seen by others who know him well as behaving in a more mature fashion.

Not only are these changes shown by various studies to occur during the period of therapy, but careful follow-up studies conducted six to eighteen months following the conclusion of therapy indicate that these changes persist.

Perhaps the facts I have given will make it clear why I feel that we are approaching the point where we can write a genuine equation in this subtle area of interpersonal relationships. Using all of the research findings we have, here is a tentative formulation of the crude equation which I believe contains the facts.

The more that the client perceives the therapist as real or genuine, as empathic, as having an unconditional regard for him, the more the client will move away from a static, fixed, unfeeling, impersonal type of functioning, and the more he will move toward a way of functioning marked by a fluid, changing, acceptant experiencing of differentiated personal feelings. The consequence of this movement is an alteration in personality and behavior in the direction of psychic health and maturity and more realistic relationships to self, others, and the environment.

THE SUBJECTIVE PICTURE

Up to this point I have spoken of the process of counseling and therapy objectively, stressing what we know, writing it as a crude equation in which we can at least tentatively put down the specific terms. But let me now try to approach it from the inside, and without ignoring this factual knowledge, present this equation as it occurs subjectively in both therapist and client. I want to do this because therapy in its occurrence is a highly personal, subjective experience. This experience has qualities quite different from the objective characteristics it possesses when viewed externally.

THE THERAPIST'S EXPERIENCE

To the therapist, it is a new venture in relating. He feels, "Here is this other person, my client. I'm a little afraid of him, afraid of the depths in him as I am a little afraid of the depths in myself. Yet as he speaks, I begin to feel a respect for him, to feel my kinship to him. I sense how frightening his world is for him, how tightly he tries to hold it in place. I would like to sense his feelings, and I would like him to know that I understand his feelings. I would like him to know that I stand with him in his tight, constricted little world, and that I can look upon it relatively unafraid. Perhaps I can make it a safer world for him. I would like my feelings in this relationship with him to be as clear and transparent as possible, so that they are a discernible reality for him, to which he can return again and again. I would like to go with him on the fearful journey into himself, into the buried fear, and hate, and love which he has never been able to let flow in him. I recognize that this is a very human and unpredictable journey for me, as well as for him, and that I may, without even knowing my fear, shrink away within myself, from some of the feelings he discovers. To this extent I know I will be limited in my ability to help him. I realize that at times his own fears may make him perceive me as uncaring, as rejecting, as an intruder, as one who does not understand. I want fully to accept these feelings in him, and yet I hope also that my own real feelings will show through so clearly that in time he cannot fail to perceive them. Most of all I want him to encounter in me a real person. I do not need to be uneasy as to whether my own feelings are 'therapeutic.' What I am and what I feel are good enough to be a basis for therapy, if I can transparently *be* what I am and what I feel in relationship to him. Then perhaps he can be what he is, openly and without fear."

The Client's Experience

And the client, for his part, goes through far more complex sequences which can only be suggested. Perhaps schematically his feelings change in some of these ways. "I'm afraid of him. I want help, but I don't know whether to trust him. He might see things which I don't know in myself — frightening and bad elements. He seems not to be judging me, but I'm sure he is. I can't tell him what really concerns me, but I can tell him about some past experiences which are related to my concern. He seems to understand those, so I can reveal a bit more of myself.

"But now that I've shared with him some of this bad side of me, he despises me. I'm sure of it, but it's strange I can find little evidence of it. Do you suppose that what I've told him isn't so bad? Is it possible that I need not be ashamed of it as a part of me? I no longer feel that he despises me. It makes me feel that I want to go further, exploring *me*, perhaps expressing more of myself. I find him a sort of companion as I do this — he seems really to understand.

"But now I'm getting frightened again, and this time deeply frightened. I didn't realize that exploring the unknown recesses of myself would make me feel feelings I've never experienced before. It's very strange because in one way these aren't new feelings. I sense that they've always been there. But they seem so bad and disturbing I've never dared to let them flow in me. And now as I live these feelings in the hours with him, I feel terribly shaky, as though my world is falling apart. It used to be sure and firm. Now it is loose, permeable and vulnerable. It isn't pleasant to feel things I've always been frightened of before. It's his fault. Yet curiously I'm eager to see him and I feel more safe when I'm with him.

"I don't know who I am any more, but sometimes when I *feel* things I seem solid and real for a moment. I'm troubled by the contradictions I find in myself — I act one way and feel another — I think one thing and feel another. It is very disconcerting. It's also sometimes adventurous and exhilarating to be trying to discover who I am. Sometimes I catch myself feeling that perhaps the person I am is worth being, whatever that means.

"I'm beginning to find it very satisfying, though often painful, to share just what it is I'm feeling at this moment. You know, it is really helpful to try to listen to myself, to hear what is going on in me. I'm not so frightened any more of what *is* going on in me. It seems pretty trust-worthy. I use some of my hours with him to dig deep into myself to know what I *am* feeling. It's scary work, but I want to *know*. And I do trust him most of the time, and that helps. I feel pretty vulnerable and raw, but I know he doesn't want to hurt me, and I even believe he cares. It occurs to me as I try to let myself down and down, deep into myself, that maybe if I could sense what is going on in me, and could realize its meaning, I would know who I am, and I would also know what to do. At least I feel this knowing sometimes with him.

"I can even tell him just how I'm feeling toward him at any given moment and instead of this killing the relationship, as I used to fear, it seems to deepen it. Do you suppose I could be my feelings with other people also? Perhaps that wouldn't be too dangerous either.

"You know, I feel as if I'm floating along on the current of life, very adventurously, being me. I get defeated sometimes, I get hurt sometimes, but I'm learning that those experiences are not fatal. I don't *know* exactly *who* I am, but I can feel my reactions at any given moment, and they seem to work out pretty well as a basis for my behavior from moment to moment. Maybe this is what it *means* to be *me*. But of course I can only do this because I feel safe in the relationship with my therapist. Or could I be myself this way outside of this relationship? I wonder. I wonder. Perhaps I could."

What I have just presented doesn't happen rapidly. It may take years. It may not, for reasons we do not understand very well, happen at all. But at least this may suggest an inside view of the factual picture I have tried to present of the process of psychotherapy as it occurs in both the therapist and his client.

PART III

The Process of Becoming a Person

I have observed the process by which an individual grows and changes in a therapeutic relationship.

5

Some of the Directions Evident in Therapy

In Part II, although there are some brief descriptions of the process of change in the client, the major focus was on the relationship which makes these changes possible. In this and the following chapter, the material deals in a much more specific way with the nature of the client's experience of change in himself.

I have a personal fondness for this chapter. It was written in 1951–52, at a time when I was making a real effort to let myself sense, and then express, the phenomena which seemed central to therapy. My book, Client-Centered Therapy, *had just been published, but I was already dissatisfied with the chapter on the process of therapy, which had of course been written about two years previously. I wanted to find a more dynamic way of communicating what happens to the person.*

So I took the case of one client whose therapy had had much significance for me, one which I was also studying from a research point of view, and using this as a basis, tried to express the tentative perceptions of the therapeutic process which were emerging in me. I felt very bold, and very unsure of myself, in pointing out that in successful therapy clients seem to come to have real affection for themselves. I felt even more uncertain in voicing the hypothesis that the core of man's nature is essentially positive. I could not then fore-

see that both of these points would receive increasing support from my experience.

THE PROCESS OF PSYCHOTHERAPY, as we have come to know it from a client-centered orientation, is a unique and dynamic experience, different for each individual, yet exhibiting a lawfulness and order which is astonishing in its generality. As I have become increasingly impressed by the inevitability of many aspects of this process, I likewise grow increasingly annoyed at the type of questions which are so commonly raised in regard to it: "Will it cure a compulsion neurosis?" "Surely you don't claim that it will erase a basic psychotic condition?" "Is it suitable for dealing with marital problems?" "Does it apply to stutterers or homosexuals?" "Are the cures permanent?" These questions, and others like them, are understandable and legitimate just as it would be reasonable to inquire whether gamma rays would be an appropriate cure for chilblains. They are however, it seems to me, the wrong questions to ask if we are trying to further a deep knowledge of what psychotherapy is, or what it may accomplish. In this chapter I should like to ask what appears to me a sounder question in regard to this fascinating and lawful process we term therapy, and to attempt a partial answer.

Let me introduce my question in this way. Whether by chance, by insightful understanding, by scientific knowledge, by artistry in human relaionships, or by a combination of all of these elements, we have learned how to initiate a describable process which appears to have a core of sequential, orderly events, which tend to be similar from one client to another. We know at least something of the attitudinal conditions for getting this process under way. We know that if the therapist holds within himself attitudes of deep respect and full acceptance for this client as he is, and similar attitudes toward the client's potentialities for dealing with himself and his situations; if these attitudes are suffused with a sufficient warmth, which transforms them into the most profound type of liking or affection for the core of the person; and if a level of communication is reached so that the client can begin to perceive that the therapist understands the feelings he is experiencing and accepts him at the full depth of that understanding, then we may be sure that the process is already initiated. Then, instead of trying to insist that this process serve the ends we have in mind (no matter how laudable those goals may be), let us ask the only question by which science can genuinely be advanced. This question is: "What is the nature of this process, what seem to be its inherent characteristics, what direction or directions does it take, and what, if any, are the natural end-points of the process?" When Benjamin Franklin observed the spark coming from the key on his kite-string, he did not, fortunately, fall under the spell of its immediate and practical uses. Instead, he began to inquire into the basic process which made such a phenomenon possible. Though many of the answers which were put forward were full of specific errors, the search was fruitful, because the right question was being asked. Thus I am making a plea that we ask the same question of psychotherapy, and ask it with open mind — that we endeavor to describe, study, and understand the basic process which underlies therapy, rather than attempting to warp that process to fit our clinical needs, or our preconceived dogma, or the evidence from some other field. Let us patiently examine it for what it *is*, in *itself*.

I have recently made an attempt to begin such a description of client-centered therapy (3). I will not repeat this description here, except to say that from the clinical and research evidence there seem to emerge certain persistent characteristics in the process: the increase in insightful statements, in maturity of reported behavior, in positive attitudes, as therapy progresses; the changes in perception of, and acceptance of, the self; the incorporation of previously denied experience into the self-structure; the shift in the locus of evaluation from outside to inside the self; the changes in the therapeutic relationship; and characteristic changes in personality structure, in behavior, and in physiological condition. Faulty as some of these descriptions may prove to be, they are an attempt to understand the process of client-centered therapy in its own terms, as revealed in clinical experience, in electrically recorded verbatim cases, and in the forty or more research studies which have been completed in this area.

My purpose in this paper is to push out beyond this material and to formulate certain trends in therapy which have received less emphasis. I should like to describe some of the directions and end points which appear to be inherent in the therapeutic process, which we have only recently begun to discern with clarity, which seem to represent significant learnings, and on which research is, as yet, nonexistent. In an attempt to convey meanings more adequately I shall use illustrative material from recorded interviews from one case. I shall also limit my discussion to the process of client-centered therapy since I have reluctantly come to concede the possibility that the process, directions, and end points of therapy may differ in different therapeutic orientations.

THE EXPERIENCING OF THE POTENTIAL SELF

One aspect of the process of therapy which is evident in all cases, might be termed the awareness of experience, or even "the experiencing of experience." I have here labelled it as the experiencing of the self, though this also falls short of being an accurate term. In the security of the relationship with a client-centered therapist, in the absence of any actual or implied threat to self, the client can let himself examine various aspects of his experience as they actually feel to him, as they are apprehended through his sensory and visceral equipment, without distorting them to fit the existing concept of self. Many of these prove to be in extreme contradiction to the concept of self, and could not ordinarily be experienced in their fullness, but in this safe relationship they can be permitted to seep through into awareness without distortion. Thus they often follow the schematic pattern, "I am thus and so, but I experience this feeling which is very inconsistent with what I am"; "I love my parents, but I experience some surprising bitterness toward them at times"; "I am really no good, but sometimes I seem to feel that I'm better than everyone else." Thus at first the expression is that "I am a self which is different from a part of my experience." Later this changes to the tentative pattern, "Perhaps I am several quite different selves, or perhaps my self contains more contradictions than I had dreamed." Still later the pattern changes to some such pattern as this: "I was sure that I could not be my experience — it was too contradictory — but now I am beginning to believe that I can be *all* of my experience."

Perhaps something of the nature of this aspect of therapy may be conveyed from two excerpts from the case of Mrs. Oak. Mrs. Oak was a housewife in her late thirties, who was having difficulties in marital and family relationships when she came in for therapy. Unlike many clients, she had a keen and spontaneous interest in the processes which she felt going on within herself, and her recorded interviews contain much material, from her own frame of reference, as to her perception of what is occurring. She thus tends to put into words what seems to be implicit, but unverbalized, in many clients. For this reason, most of the excerpts in this chapter will be taken from this one case.

From an early portion of the fifth interview comes material which describes the awareness of experience which we have been discussing.

Client: It all comes pretty vague. But you know I keep, keep having the thought occur to me that this whole process for me is kind of like examining pieces of a jig-saw puzzle. It seems to me I, I'm in the process now of examining the individual pieces which really don't have too much meaning. Probably handling them, not even beginning to think of a pattern. That keeps coming to me. And it's interesting to me because I, I really don't like jig-saw puzzles. They've always irritated me. But that's my feeling. And I mean I pick up little pieces (*she gestures throughout this conversation to illustrate her statements*) with absolutely no meaning except I mean the, the feeling that you get from simply handling them without seeing them as a pattern, but just from the touch, I probably feel, well it is going to fit someplace here.

Therapist: And that at the moment that, that's the process, just getting the feel and the shape and the configuration of the different pieces with a little bit of background feeling of, yeah they'll probably fit somewhere, but most of the attention's focused right on, "What does this feel like? And what's its texture?"

C: That's right. There's almost something physical in it. A, a —

T: You can't quite describe it without using your hands. A real, almost a sensuous sense in —

C: That's right. Again it's, it's a feeling of being very objective, and yet I've never been quite so close to myself.

T: Almost at one and the same time standing off and looking at yourself and yet somehow being closer to yourself that way than —

C: M-hm. And yet for the first time in months I am not thinking about my problems. I'm not actually, I'm not working on them.

T: I get the impression you don't sort of sit down to work on "my problems." It isn't that feeling at all.

C: That's right. That's right. I suppose what I, I mean actually is that I'm not sitting down to put this puzzle together as, as something, I've got to see the picture. It, it may be that, it may be that I am actually enjoying this feeling process. Or I'm certainly learning something.

T: At least there's a sense of the immediate goal of getting that feel as being the thing, not that you're doing this in order to see a picture, but that it's a, a satisfaction of really getting acquainted with each piece. Is that —

C: That's it. That's it. And it still becomes that sort of sensuousness, that touching. It's quite interesting. Sometimes not entirely pleasant, I'm sure, but —

T: A rather different sort of experience.

C: Yes. Quite.

This excerpt indicates very clearly the letting of material come into awareness, without any attempt to own it as part of the self, or to relate it to other material held in consciousness. It is, to put it as accurately as possible, an awareness of a wide range of experiences, with, at the moment, no thought of their relation to self. Later it may be recognized that what was being experienced may all become a part of self. Thus the heading of this section has been termed "The Experiencing of the Potential Self."

The fact that this is a new and unusual form of experience is expressed in a verbally confused but emotionally clear portion of the sixth interview.

C: Uh, I caught myself thinking that during these sessions, uh, I've been sort of singing a song. Now that sounds vague and uh — not actually singing — sort of a song without any music. Probably a kind of poem coming out. And I like the idea, I mean it's just sort of come to me without anything built out of, of anything. And in — following that, it came, it came this other kind of feeling. Well, I found myself sort of asking myself, is that the shape that cases take? Is it possible that I am just verbalizing and, at times kind of become intoxicated with my own verbalizations? And then uh, following this, came, well, am I just taking up your time? And then a doubt, a doubt. Then something else occurred to me. Uh, from whence it came, I don't know, no actual logical kind of sequence to the thinking. The thought struck me: We're doing bits, uh, we're not overwhelmed or doubtful, or show concern or, or any great interest when, when blind people learn to read with their fingers, Braille. I don't know — it may be just sort of, it's all mixed up. It may be that's something that I'm experiencing now.

T: Let's see if I can get some of that, that sequence of feelings. First, sort of as though you're, and I gather that first one is a fairly positive feeling, as though maybe you're kind of creating a poem here — a song without music somehow but something that might be quite creative, and then the, the feeling of a lot of skepticism about that. "Maybe I'm just saying words, just being carried off by words that I, that I speak, and maybe it's all a lot of baloney, really." And then a feeling that perhaps you're almost learning a new type of experiencing which would be just as radically new as for a blind person to try to make sense out of what he feels with his fingertips.

C: M-hm. M-hm. (*Pause*) . . . And I sometimes think to myself, well, maybe we could go into this particular incident or that particular incident. And then somehow when I come here, there is, that doesn't hold true, it's, it seems false. And then there just seems to be this flow of words which somehow aren't forced and then occasionally this doubt creeps in. Well, it sort of takes form of a, maybe you're just making music. . . . Perhaps that's why I'm doubtful today of, of this whole thing, because it's something that's not forced. And really I'm feeling that what I should do is, is sort of systematize the thing. Oughta work harder and —

T: Sort of a deep questioning as to what am I doing with a self that isn't, isn't pushing to get things *done, solved?* (*Pause*)

C: And yet the fact that I, I really like this other kind of thing, this, I don't know, call it a poignant feeling, I mean — I felt things that I never felt before. I *like* that, too. Maybe that's the way to do it. I just don't know today.

Here is the shift which seems almost invariably to occur in therapy which has any depth. It may be represented schematically as the client's feeling that "I came here to solve problems, and now I find myself just experiencing myself." And as with this client this shift is usually accompanied by the intellectual formulation that it is wrong, and by an emotional appreciation of the fact that it "feels good."

We may conclude this section saying that one of the fundamental directions taken by the process of therapy is the free experiencing of the actual sensory and visceral reactions of the organism without too much of an attempt to relate these experiences to the self. This is usually accompanied by the conviction that this material does not belong to, and cannot be organized into, the self. The end point of this process is that the client discovers that he can *be* his experience, with all of its variety and surface contradiction; that he can formulate himself out of his experience, instead of trying to impose a formulation of self upon his experience, denying to awareness those elements which do not fit.

The Full Experiencing of an Affectional Relationship

One of the elements in therapy of which we have more recently become aware is the extent to which therapy is a learning, on the part of the client, to accept fully and freely and without fear the positive feelings of another. This is not a phenomenon which clearly occurs in every case. It seems particularly true of our longer cases, but does not occur uniformly in these. Yet it is such a deep experience that we have begun to question whether it is not a highly significant direction in the therapeutic process, perhaps occurring at an unverbalized level to some degree in all successful cases. Before discussing this phenomenon, let us give it some body by citing the experience of Mrs. Oak. The experience struck her rather suddenly, between the twenty-ninth and thirtieth interview, and she spends most of the latter interview discussing it. She opens the thirtieth hour in his way.

> *C:* Well, I made a very remarkable discovery. I know it's — (*laughs*) I found out that you actually *care* how this thing goes. (*Both laugh*) It gave me the feeling, it's sort of well — "maybe I'll let you get in the act," sort of thing. It's — again you see, on an examination sheet, I would have had the correct answer, I mean — but it suddenly dawned on me that in the — client-counselor kind of thing, you *actually care* what happens to this thing. And it was a revelation, a — not that. That doesn't describe it. It was a — well, the closest I can come to it is a kind of relaxation, a — not a letting down, but a — (*pause*) more of a straightening out without tension if that means anything. I don't know.
>
> *T:* Sounds as though it isn't as though this was a new idea, but it was a new *experience* of really *feeling* that I did care and if I get the rest of that, sort of a willingness on your part to let me care.
>
> *C:* Yes.

This letting the counselor and his warm interest into her life was undoubtedly one of the deepest features of therapy in this case. In an interview following the conclusion of therapy she spontaneously mentions this experience as being the outstanding one. What does it mean?

The phenomenon is most certainly not one of transference and countertransference. Some experienced psychologists who had undergone psychoanalysis had the opportunity of observing the development of the relationship in another case than the one cited. They were the first to object to the use of the terms transference and countertransference to describe the phenomena. The gist of their remarks was that this is something which is mutual and appropriate, where transference or countertransference are phenomena which are characteristically one-way and inappropriate to the realities of the situation.

Certainly one reason why this phenomena is occurring more frequently in our experience is that as therapists we have become less afraid of our positive (or negative) feelings toward the client. As therapy goes on the therapist's feeling of acceptance and respect for the client tends to change to something approaching awe as he sees the valiant and deep struggle of the person to be himself. There is, I think, within the therapist, a profound experience of the underlying commonality — should we say brotherhood — of man. As a result he feels toward the client a warm, positive, affectional reaction. This poses a problem for the client who often, as in this case, finds it difficult to accept the positive feeling of another. Yet once accepted the inevitable reaction on the part of the client is to relax, to let the warmth of liking by another person reduce the tension and fear involved in facing life.

But we are getting ahead of our client. Let us examine some of the other aspects of this experience as it occurred to her. In earlier interviews she had talked of the fact that she did *not* love humanity, and that in some vague and stubborn way she felt she was right, even though others would regard her as wrong. She mentions this again as she discusses the way this experience has clarified her attitudes toward others.

> *C:* The next thing that occurred to me that I found myself thinking and still thinking, is somehow — and I'm not clear why — the same kind of a caring that I get when I say "I don't love humanity." Which has always sort of — I mean I was always convinced of it. So I mean, it doesn't — I knew that it was a good thing, see. And I think I clarified it within myself — what it has to do with this situation, I don't know. But I found out, no, I don't love, but I do *care* terribly.

T: M-hm. M-hm. I see. . . .

C: . . . It might be expressed better in saying I care terribly what happens. But the caring is a — takes form — its structure is in understanding and not wanting to be taken in, or to contribute to those things which I feel are false and — It seems to me that in — in loving, there's a kind of *final* factor. If you do that, you've sort of done *enough.* It's a —

T: That's *it,* sort of.

C: Yeah. It seems to me this other thing, this caring, which isn't a good term — I mean, probably we need something else to describe this kind of thing. To say it's an impersonal thing doesn't mean anything because it isn't impersonal. I mean I feel it's very much a part of a whole. But it's something that somehow doesn't stop. . . . It seems to me you could have this feeling of loving humanity, loving people, and at the same time — go on contributing to the factors that make people neurotic, make them ill — where, what I feel is a resistance to those things.

T: You care enough to want to understand and to want to avoid contributing to anything that would make for more neuroticism, or more of that aspect in human life.

C: Yes. And it's — (*pause*). Yes, it's something along those lines. . . . Well, again, I have to go back to how I feel about this other thing. It's — I'm not really called upon to give of myself in a — sort of on the auction block. There's nothing final. . . . It sometimes bothered me when I — I would have to say to myself, "I don't love humanity," and yet, I always knew that there was something positive. That I was probably right. And — I may be all off the beam now, but it seems to me that, that is somehow tied up in the — this feeling that I — I have now, into how the therapeutic value can carry through. Now, I couldn't tie it up, I couldn't tie it in, but it's as close as I can come to explaining to myself, my — well, shall I say the learning process, the follow through on my realization that — yes, you *do care* in a given situation. It's just that simple. And I hadn't been aware of it before. I might have closed this door and walked out, and in discussing therapy, said, yes, the counselor must feel thus and so, but, I mean, I hadn't had the dynamic experience.

In this portion, though she is struggling to describe her own feeling, it would seem that what she is saying would be characteristic of the therapist's attitude toward the client as well. His attitude, at its best, is devoid of the *quid pro quo* aspect of most of the experiences we call love. It is the simple outgoing human feeling of one individual for another, a feeling, it seems to me which is even more basic than sexual or parental feeling. It is a caring enough about the person that you do not wish to interfere with his development, nor to use him for any self-aggrandizing goals of your own. Your satisfaction comes in having set him free to grow in his own fashion.

Our client goes on to discuss how hard it has been for her in the past to accept any help or positive feeling from others, and how this attitude is changing.

C: I have a feeling . . . that you have to do it pretty much yourself, but that somehow you ought to be able to do that with other people. (*She mentions that there have been "countless" times when she might have accepted personal warmth and kindliness from others.*) I get the feeling that I just was afraid I would be devastated. (*She returns to talking about the counseling itself and her feeling toward it.*) I mean there's been this tearing through the thing myself. Almost to — I mean, I felt it — I mean I tried to verbalize it on occasion — a kind of — at times almost not wanting you to restate, not wanting you to reflect, the thing is *mine*. Course all right, I can say it's resistance. But that doesn't mean a damn thing to me now. . . . The — I think in — in relationship to this particular thing, I mean, the — probably at times, the strongest feeling was, it's mine, it's *mine*. I've got to cut it down myself. See?

T: It's an experience that's awfully hard to put down accurately into words, and yet I get a sense of difference here in this relationship, that from the feeling that "this is mine," "I've got to do it," "I am doing it," and so on, to a somewhat different feeling that — "I could let you in."

C: Yeah. Now. I mean, that's — that it's — well, it's sort of, shall we say, volume two. It's — it's a — well, sort of, well. I'm still in the thing alone, but I'm *not* — see — I'm —

T: M-hm. Yes, that paradox sort of sums it up, doesn't it?

C: Yeah.

T: In all of this, there is a feeling, it's still — every aspect of my experience is mine and that's kind of inevitable and necessary and so on. And yet that isn't the whole picture either. Somehow it can be shared or another's interest can come in and in some ways it is new.

C: Yeah. And it's — it's as though, that's how it should be. I mean, that's how it — has to be. There's a — there's a feeling, "and this is good." I mean, it expresses, it clarifies it for me. There's a feeling — in this caring, as though — you were sort of standing back — standing off, and if I want to sort of cut through to the thing, it's a — a slashing of — oh, tall weeds, that I can do it, and you can — I mean you're not going to be disturbed by having to walk through it, too. I don't know. And it doesn't make sense. I mean —

T: Except there's a very real sense of rightness about this feeling that you have, hm?

C: M-hm.

May it not be that this excerpt portrays the heart of the process of socialization? To discover that it is *not* devastating to accept the positive feeling from another, that it does not necessarily end in hurt, that it actually "feels good" to have another person with you in your struggles to meet life — this may be one of the most profound learnings encountered by the individual whether in therapy or not.

Something of the newness, the non-verbal level of this experience is described by Mrs. Oak in the closing moments of this thirtieth interview.

C: I'm experiencing a new type, a — probably the only worthwhile kind of learning, a — I know I've — I've often said what I know doesn't help me here. What I meant is, my acquired knowledge doesn't help me. But it seems to me that the learning process here has been — so dynamic, I mean, so much a part of the — of everything, I mean, of me, that if I just get that out of it, it's something, which, I mean — I'm wondering if I'll ever be able to straighten out into a sort of acquired knowledge what I have experienced here.

T: In other words, the kind of learning that has gone on here has been something of quite a different sort and quite a different depth; very vital, very real. And quite worthwhile to you in and of itself, but the question you're asking is: Will I ever have a clear intellectual picture of what has gone on at this somehow deeper kind of learning level?

C: M-hm. Something like that.

Those who would apply to therapy the so-called laws of learning derived from the memorization of nonsense syllables would do well to study this excerpt with care. Learning as it takes place in therapy is a total, organismic, frequently non-verbal type of thing which may or may not follow the same principles as the intellectual learning of trivial material which has little relevance to the self. This, however, is a digression.

Let us conclude this section by rephrasing its essence. It appears possible that one of the characteristics of deep or significant therapy is that the client discovers that it is not devastating to admit fully into his own experience the positive feeling which another, the therapist, holds toward him. Perhaps one of the reasons why this is so difficult is that essentially it involves the feeling that "I am worthy of being liked." This we shall consider in the following section. For the present it may be pointed out that this aspect of therapy is a free and full experiencing of an affectional relationship which may be put in generalized terms as follows: "I can permit someone to care about me, and can fully accept that caring within myself. This permits me to recognize that I care, and care deeply, for and about others."

The Liking of One's Self

In various writings and researches that have been published regarding client-centered therapy there has been a stress upon the acceptance of self as one of the directions and outcomes of therapy. We have established the fact that in successful psychotherapy negative attitudes toward the self decrease and positive attitudes increase. We have measured the gradual increase in self-acceptance and have studied the correlated increase in acceptance of others. But as I examine these statements and compare them with our more recent cases, I feel they fall short of the truth. The client not only accepts himself — a phrase which may carry the connotation of a grudging and reluctant acceptance of the inevitable — he actually comes to *like* himself. This is not a bragging or self-assertive liking; it is rather a quiet pleasure in being one's self.

Mrs. Oak illustrates this trend rather nicely in her thirty-third interview. Is it significant that this follows by ten days the interview where she could for the first time admit to herself that the therapist cared? Whatever our speculations on this point, this fragment indicates very well the quiet joy in being one's self, together with the apologetic attitude which, in our culture, one feels it is necessary to take toward such an experience. In the last few minutes of the interview, knowing her time is nearly up she says:

C: One thing worries me — and I'll hurry because I can always go back to it — a feeling that occasionally I can't turn out. Feeling of being quite pleased with myself. Again the Q technique.* I walked out of here one time, and impulsively I threw my first card, "I am an attractive personality"; looked at it sort of aghast but left it there, I mean, because honestly, I mean, that is exactly how it felt — a — well, that bothered me and I catch that now.

* This portion needs explanation. As part of a research study by another staff member this client had been asked several times during therapy to sort a large group of cards, each containing a self-descriptive phrase, in such a way as to portray her own self. At one end of the sorting she was to place the card or cards most like herself, and at the other end, those most unlike herself. Thus when she says that she put as the first card, "I am an attractive personality," it means that she regarded this as the item most characteristic of herself.

Every once in a while a sort of pleased feeling, nothing superior, but just — I don't know, sort of pleased. A neatly turned way. And it bothered me. And yet — I wonder — I rarely remember things I say here, I mean I wondered why it was that I was convinced, and something about what I've felt about being hurt that I suspected in — my feelings when I would hear someone say to a child, "Don't cry." I mean, I always felt, but it isn't right; I mean, if he's hurt, let him cry. Well, then, now this pleased feeling that I have. I've recently come to feel, it's — there's something almost the same there. It's — We don't object when *children* feel pleased with themselves. It's — I mean, there really isn't anything vain. It's — maybe that's how people *should* feel.

T: You've been inclined almost to look askance at yourself for this feeling, and yet as you think about it more, maybe it comes close to the two sides of the picture, that if a child wants to cry, why shouldn't he cry? And if he wants to feel pleased with himself, doesn't he have a perfect right to feel pleased with himself? And that sort of ties in with this, what I would see as an appreciation of yourself that you've experienced every now and again.

C: Yes. Yes.

T: "I'm really a pretty rich and interesting person."

C: Something like that. And then I say to myself, "Our society pushes us around and we've lost it." And I keep going back to my feelings about children. Well, maybe they're richer than we are. Maybe we — it's something we've lost in the process of growing up.

T: Could be that they have a wisdom about that that we've lost.

C: That's right. My time's up.

Here she arrives, as do so many other clients, at the tentative, slightly apologetic realization that she has come to like, enjoy, appreciate herself. One gets the feeling of a spontaneous relaxed enjoyment, a primitive *joie de vivre,* perhaps analogous to the lamb frisking about the meadow or the porpoise gracefully leaping in and out of the waves. Mrs. Oak feels that it is something native to the organism, to the infant, something we have lost in the warping process of development.

Earlier in this case one sees something of a forerunner of this feeling, an incident which perhaps makes more clear its fundamental nature. In the ninth interview Mrs. Oak in a somewhat embarrassed fashion reveals something she has always kept to herself. That she brought it forth at some cost is indicated by the fact that it was preceded by a very long pause, of several minutes duration. Then she spoke.

C: You know this is kind of goofy, but I've never told anyone this (*nervous laugh*) and it'll probably do me good. For years, oh, probably from early youth, from seventeen probably on, I, I have had what I have come to call to myself, told myself were "flashes of sanity." I've never told anyone this, (*another embarrassed laugh*) wherein, in, really I feel sane. And, and pretty much aware of life. And always with a terrific kind of concern and sadness of how far away, how far astray that we have actually gone. It's just a feeling once in a while of finding myself a whole kind of person in a terribly chaotic kind of world.

T: It's been fleeting and it's been infrequent, but there have been times when it seems the whole you is functioning and feeling in the world, a very chaotic world to be sure —

C: That's right. And I mean, and knowing actually how far astray we, we've gone from, from being whole healthy people. And of course, one doesn't talk in those terms.

T: A feeling that it wouldn't be *safe* to talk about the singing you* —

C: Where does that person live?

T: Almost as if there was no place for such a person to, to exist.

C: Of course, you know, that, that makes me — now wait a minute — that probably explains why I'm primarily concerned with feelings here. That's probably it.

* The therapist's reference is to her statement in a previous interview that in therapy she was singing a song.

T: Because that whole you does exist with all your feelings. Is that it, you're more aware of feelings?

C: That's right. It's not, it doesn't reject feelings and — that's *it.*

T: That whole you somehow lives feelings instead of somehow pushing them to one side.

C: That's right. (*Pause*) I suppose from the practical point of view it could be said that what I ought to be doing is solving some problems, day-to-day problems. And yet, I, I — what I'm trying to do is solve, solve something else that's a great, that is a great deal more important than little day-to-day problems. Maybe that sums up the whole thing.

T: I wonder if this will distort your meaning, that from a hard-headed point of view you ought to be spending time thinking through specific problems. But you wonder if perhaps maybe you aren't on a quest for this whole you and perhaps that's more important than a solution to the day-to-day problems.

C: I think that's it. I think that's it. That's probably what I mean.

If we may legitimately put together these two experiences, and if we are justified in regarding them as typical, then we may say that both in therapy and in some fleeting experiences throughout her previous life, she has experienced a healthy satisfying enjoyable appreciation of herself as a whole and functioning creature; and that this experience occurs when she does not reject her feelings but lives them.

Here it seems to me is an important and often overlooked truth about the therapeutic process. It works in the direction of permitting the person to experience fully, and in awareness, all of his reactions including his feelings and emotions. As this occurs, the individual feels a positive liking for himself, a genuine appreciation of himself as a total functioning unit, which is one of the important end points of therapy.

THE DISCOVERY THAT THE CORE OF PERSONALITY IS POSITIVE

One of the most revolutionary concepts to grow out of our clinical experience is the growing recognition that the innermost core of man's nature, the deepest layers of his personality, the base of his "animal nature," is positive in nature — is basically socialized, forward-moving, rational and realistic.

This point of view is so foreign to our present culture that I do not expect it to be accepted, and it is indeed so revolutionary in its implications that it should not be accepted without thorough-going inquiry. But even if it should stand these tests, it will be difficult to accept. Religion, especially the Protestant Christian tradition, has permeated our culture with the concept that man is basically sinful, and only by something approaching a miracle can his sinful nature be negated. In psychology, Freud and his followers have presented convincing arguments that the id, man's basic and unconscious nature, is primarily made up of instincts which would, if permitted expression, result in incest, murder, and other crimes. The whole problem of therapy, as seen by this group, is how to hold these untamed forces in check in a wholesome and constructive manner, rather than in the costly fashion of the neurotic. But the fact that at heart man is irrational, unsocialized, destructive of others and self — this is a concept accepted almost without question. To be sure there are occasional voices of protest. Maslow (1) puts up a vigorous case for man's animal nature, pointing out that the anti-social emotions — hostility, jealousy, etc. — result from frustration of more basic impulses for love and security and belonging, which are in themselves desirable. And Montagu (2) likewise develops the thesis that cooperation, rather than struggle, is the basic law of human life. But these solitary voices are little heard. On the whole the viewpoint of the professional worker as well as the layman is that man as he is, in his basic nature, had best be kept under control or under cover or both.

As I look back over my years of clinical experience and research, it seems to me that I have been very slow to recognize the falseness of this popular and professional concept. The reason, I believe, lies in the fact that in therapy there are continually being uncovered hostile and anti-social feelings, so that it is easy to assume that this indicates the deeper and therefore the basic nature of man. Only slowly has it become evident that these untamed and unsocial feelings are neither the deepest nor the strongest, and that the inner core of man's personality is the organism itself, which is essentially both self-preserving and social.

To give more specific meaning to this argument, let me turn again to the case of Mrs. Oak. Since the point is an important one, I shall quote at some length from the recorded case to illustrate the type of experience on which I have based the foregoing statements. Perhaps the excerpts can illustrate the opening up of layer after layer of personality until we come to the deepest elements.

It is in the eighth interview that Mrs. Oak rolls back the first layer of defense, and discovers a bitterness and desire for revenge underneath.

C: You know over in this area of, of sexual disturbance, I have a feeling that I'm beginning to discover that it's pretty bad, pretty bad. I'm finding out that, that I'm bitter, really. Damn bitter. I — and I'm not turning it back in, into myself . . . I think what I probably feel is a certain element of "I've been cheated." (*Her voice is very tight and her throat chokes up.*) And I've covered up very nicely, to the point of consciously not caring. But I'm, I'm sort of amazed to find that in this practice of, what shall I call it, a kind of sublimation that right under it — again words — there's a, a kind of passive force that's, it's pas — it's very passive, but at the same time it's just kind of *murderous.*

T: So there's the feeling, "I've really been cheated. I've covered that up and seem not to care and yet underneath that there's a kind of a, a latent but very much present *bitterness* that is very, very strong."

C: It's very strong. I — that I know. It's terribly powerful.

T: Almost a dominating kind of force.

C: Of which I am rarely conscious. Almost never . . . Well, the only way I can describe it, it's a kind of murderous thing, but without violence. . . . It's more like a feeling of wanting to get even. . . . And of course, I won't pay back, but I'd like to. I really would like to.

Up to this point the usual explanation seems to fit perfectly. Mrs. Oak has been able to look beneath the socially controlled surface of her behavior, and finds underneath a murderous feeling of hatred and a desire to get even. This is as far as she goes in exploring this particular feeling until considerably later in therapy. She picks up the theme in the thirty-first interview. She has had a hard time getting under way, feels emotionally blocked, and cannot get at the feeling which is welling up in her.

C: I have the feeling it isn't guilt. (*Pause. She weeps.*) Of course I mean, I can't verbalize it yet. (*Then with a rush of emotion*) It's just being *terribly hurt!*

T: M-hm. It isn't guilt except in the sense of being very much wounded somehow.

C: (*Weeping*) It's — you know, often I've been guilty of it myself but in later years when I've heard parents say to their children, "stop crying," I've had a feeling, a hurt as though, well, why should they tell them to stop crying? They feel sorry for themselves, and who can feel more adequately sorry for himself than the child. Well, that is sort of what — I mean, as though I mean, I thought that they should let him cry. And — feel sorry for him too, maybe. In a rather objective kind of way. Well, that's — that's something of the kind of thing I've been experiencing. I mean, now — just right now. And in — in —

T: That catches a little more the flavor of the feeling that it's almost as if you're really weeping for yourself.

C: Yeah. And again you see there's conflict. Our culture is such that — I mean, one doesn't indulge in self-pity. But this isn't — I mean, I feel it doesn't quite have that connotation. It may have.

T: Sort of think that there is a cultural objection to feeling sorry about yourself. And yet you feel the feeling you're experiencing isn't quite what the culture objected to either.

C: And then of course, I've come to — to see and to feel that over this — see, I've covered it up. (*Weeps.*) But I've covered it up with so much *bitterness,* which in turn I had to cover up. (*Weeping*) *That's* what I want to get rid of! I almost don't *care* if I hurt.

T: (*Softly, and with an empathic tenderness toward the hurt she is experiencing*) You feel that here at the basis of it as you experience it is a feeling of real tears for yourself. But *that* you can't show, mustn't show, so that's been covered by bitterness that you don't like, that you'd like to be rid of. You almost feel you'd rather absorb the hurt than to — than to feel the bitterness. (*Pause*) And what you seem to be saying quite strongly is, I do *hurt,* and I've tried to cover it up.

C: I didn't *know* it.

T: M-hm. Like a new discovery really.

C: (*Speaking at the same time*) I never really did know. But it's — you know, it's almost a physical thing. It's — it's sort of as though I were looking within myself at all kinds of — nerve endings and bits of things that have been sort of mashed. (*Weeping*)

T: As though some of the most delicate aspects of you physically almost have been crushed or hurt.

C: Yes. And you know, I do get the feeling, "Oh, you poor thing." (*Pause*)

T: Just can't help but feel very deeply sorry for the person that is you.

C: I don't think I feel sorry for the whole person; it's a certain aspect of the thing.

T: Sorry to see that hurt.

C: Yeah.

T: M-hm. M-hm.

C: And then of course there's this damn bitterness that I want to get rid of. It's — it gets me into trouble. It's because it's a tricky thing. It tricks me. (*Pause*)

T: Feel as though that bitterness is something you'd like to be rid of because it doesn't do right by you.

C: (*C weeps. Long pause*) I don't know. It seems to me that I'm right in feeling, what in the world good would it do to term this thing guilt. To chase down things that would give me an interesting case history, shall we say. What *good* would it do? It seems to me that the — that the key, the real thing is in this feeling that I have.

T: You could track down some tag or other and could make quite a pursuit of that, but you feel as though the core of the whole thing is the kind of experience that you're just having right here.

C: That's right. I mean if — I don't know what'll happen to the feeling. Maybe nothing. I don't know, but it seems to me that whatever understanding I'm to have is a part of this feeling of hurt, of — it doesn't matter much what it's called. (*Pause*) Then I — one can't go — around with a hurt so openly exposed. I mean this seems to me that somehow the next process has to be a kind of healing.

T: Seems as though you couldn't possibly expose yourself if part of yourself is so hurt, so you wonder if somehow the hurt mustn't be healed first. (*Pause*)

C: And yet, you know, it's — it's a funny thing (*pause*). It sounds like a statement of complete confusion or the old saw that the neurotic doesn't want to give up his symptoms. But that isn't true. I mean, that isn't true here, but it's — I can just hope that this will impart what I feel. I somehow don't mind being hurt. I mean, it's just occurred to me that I don't mind terribly. It's a — I mind more the — the feeling of bitterness which is, I know, the cause of this frustration, I mean the — I somehow mind that more.

T: Would this get it? That, though you don't like the hurt, yet you feel you can accept that. That's bearable. Somehow it's the things that have covered up that hurt, like the bitterness, that you just — at this moment, can't stand.

C: Yeah. That's just about it. It's sort of as though, well, the first,

I mean, as though, it's — well, it's something I can cope with. Now, the feeling of, well, I can still have a hell of a lot of fun, see. But that this other, I mean, this frustration — I mean, it comes out in so many ways, I'm beginning to realize, you see. I mean, just this sort of, this kind of thing.

T: And a hurt you can accept. It's a part of life within a lot of other parts of life, too. You can have lots of fun. But to have all of your life diffused by frustration and bitterness, that you don't like, you don't want, and are now more aware of.

C: Yeah. And there's somehow no dodging it now. You see, I'm much more aware of it. (*Pause*) I don't know. Right now, I don't know just what the next step is. I really don't know. (*Pause*) Fortunately, this is a kind of development, so that it — doesn't carry over too acutely into — I mean, I — what I'm trying to say, I think, is that I'm still functioning. I'm still enjoying myself and —

T: Just sort of want me to know that in lots of ways you carry on just as you always have.

C: That's it. (*Pause*) Oh, I think I've got to stop and go.

In this lengthy excerpt we get a clear picture of the fact that underlying the bitterness and hatred and the desire to get back at the world which has cheated her, is a much less anti-social feeling, a deep experience of having been hurt. And it is equally clear that at this deeper level she has no desire to put her murderous feelings into action. She dislikes them and would like to be rid of them.

The next excerpt comes from the thirty-fourth interview. It is very incoherent material, as verbalizations often are when the individual is trying to express something deeply emotional. Here she is endeavoring to reach far down into herself. She states that it will be difficult to formulate.

C: I don't know whether I'll be able to talk about it yet or not. Might give it a try. Something — I mean, it's a feeling — that — sort of an urge to really get out. I know it isn't going to make sense. I think that maybe if I can get it out and get it a little, well, in a little more matter of fact way, that it'll be something that's more useful to me. And I don't know how to — I mean, it seems as though I want to say, I want to talk about my *self*. And that is of course as I see, what I've been doing for all these hours. But, no, this — it's my *self*. I've quite recently become aware of rejecting certain statements, because to me they sounded — not quite what I meant, I mean, a little bit too idealized. And I mean, I can remember always saying it's more selfish than that, more selfish than that. Until I — it sort of occurs to me, it dawns, yeah, that's exactly what I mean, but the selfishness I mean, has an entirely different connotation. I've been using a word "selfish." Then I have this feeling of — I — that I've never expressed it before, of selfish — which means nothing. A — I'm still going to talk about it. A kind of pulsation. And it's something aware all the time. And still it's there. And I'd like to be able to utilize it, too — as a kind of descending into this thing. You know, it's as though — I don't know, damn! I'd sort of acquired someplace, and picked up a kind of acquaintance with the structure. Almost as though I knew it brick for brick kind of thing. It's something that's an awareness. I mean, that — of a feeling of not being fooled, of not being drawn into the thing, and a critical sense of knowingness. But in a way — the reason, it's hidden and — can't be a part of everyday life. And there's something of — at times I feel almost a little bit terrible in the thing, but again terrible not as terrible. And why? I think I know. And it's — it also explains a lot to me. It's — it's something that is *totally* without hate. I mean, just *totally*. Not with love, but *totally without hate*. But it's — it's an exciting thing, too . . . I guess maybe I am the kind of person that likes to, I mean, probably even torment myself, or to chase things down, to try to find the whole. And I've told myself, now look, this is a pretty strong kind of feeling which you have. It isn't constant. But you feel it sometimes, and as you let yourself feel it, you feel it yourself. You know, there are words for that kind of thing that one could find in abnormal psychology. Might almost be like the feeling that is occasionally, is attributed to things that you read about. I mean, there are some elements there — I mean, this pulsation, this excitement, this knowing. And I've said — I tracked down one thing, I mean, I was very, very brave, what shall we say — a sublimated sex drive. And I thought, well, *there* I've got it. I've really solved the thing. And that there is nothing more to it than that. And for awhile, I mean, I was quite pleased with myself. That was it. And then I had to admit, no, that wasn't it. 'Cause that's something that had been with me long before I became so terribly frustrated sexually. I mean, that wasn't — and, but in the thing, then I began to see a little, within this very core is an acceptance of sexual relationship, I mean, the only kind that *I* would think would be possible. It was in this thing. It's not something that's been — I mean, sex hasn't been sublimated or substituted there. No. Within this, within what I know there — I mean, it's a different kind of sexual feeling to be sure. I mean, it's one that is stripped of all the things that have happened to sex, if you know what I mean. There's no chase, no pursuit, no battle, no — well, no kind of hate, which I think, seems to me, has crept into such things. And yet, I mean, this feeling has been, oh, a little bit disturbing.

T: I'd like to see if I can capture a little of what that means to you. It is as you've gotten very deeply acquainted with yourself on kind of a brick-by-brick experiencing basis, and in that sense have become more *self*-ish, and the notion of really, — in the discovering of what is the core of you as separate from all the other aspects, you come across the realization, which is a very deep and pretty thrilling realization, that the core of that self is not only without hate, but is really something more resembling a saint, something really very pure, is the word I would use. And that you can try to depreciate that. You can say, maybe it's a sublimation, maybe it's an abnormal manifestation, screwball and so on. But inside of yourself, you knew that it isn't. This contains the feelings which could contain rich sexual expression, but it sounds bigger than, and really deeper than that. And yet fully able to include all that could be a part of sex expression.

C: It's probably something like that. . . . It's kind of — I mean, it's a kind of descent. It's a going down where you might almost think it should be going up, but no, it's — I'm sure of it; it's kind of going down.

T: This is a going down and immersing yourself in your self almost.

C: Yeah. And I — I can't just throw it aside. I mean, it just seems, oh, it just *is.* I mean, it seems an awfully important thing that I just had to say.

T: I'd like to pick up one of those things too, to see if I understand it. That it sounds as though this sort of idea you're expressing is something you must be going up to capture, something that *isn't* quite. Actually though, the feeling is, this is a going down to capture something that's more deeply there.

C: It is. It really — there's something to that which is — I mean, this — I have a way, and of course sometime we're going to have to go into that, of rejecting almost violently, that which is righteous, rejection of the ideal, the — as — and that expressed it; I mean, that's sort of what I mean. One is a going up into I don't know. I mean, I just have a feeling, I can't follow. I mean, it's pretty thin stuff if you ever start knocking it down. This one went — I wondered why — I mean, has this awfully definite feeling of descending.

T: That this isn't a going up into the thin ideal. This is a going down into the astonishingly solid reality, that —

C: Yeah.

T: — is really more surprising than —

C: Yeah. I mean, a something that you don't knock down. That's there — I don't know — seems to me after you've abstracted the whole thing. That lasts. . . .

Since this is presented in such confused fashion, it might be worth while to draw from it the consecutive themes which she has expressed.

I'm going to talk about myself as *self*-ish, but with a new connotation to the word.

I've acquired an acquaintance with the structure of myself, know myself deeply.

As I descend into myself, I discover something exciting, a core that is totally without hate.

It can't be a part of everyday life — it may even be abnormal.

I thought first it was just a sublimated sex drive.
But no, this is more inclusive, deeper than sex.
One would expect this to be the kind of thing one would discover by going up into the thin realm of ideals.
But actually, I found it by going deep within myself.
It seems to be something that is the essence, that lasts.

Is this a mystic experience she is describing? It would seem that the counselor felt so, from the flavor of his responses. Can we attach any significance to such a Gertrude Stein kind of expression? The writer would simply point out that many clients have come to a somewhat similar conclusion about themselves, though not always expressed in such an emotional way. Even Mrs. Oak, in the following interview, the thirty-fifth, gives a clearer and more concise statement of her feeling, in a more down-to-earth way. She also explains why it was a difficult experience to face.

C: I think I'm awfully glad I found myself or brought myself or wanted to talk about self. I mean, it's a very personal, private kind of thing that you just don't talk about. I mean, I can understand my feeling of, oh, probably slight apprehension now. It's — well, sort of as though I was just rejecting, I mean, all of the things that western civilization stands for, you see. And wondering whether I was right, I mean, whether it was quite the right path, and still of course, feeling how right the thing was, you see. And so there's bound to be a conflict. And then this, and I mean, now I'm feeling, well, of course that's how I feel. I mean there's a — this thing that I term a kind of a lack of hate, I mean, is very real. It carried over into the things I do, I believe in. . . . I think it's all right. It's sort of maybe my saying to myself, well, you've been bashing me all over the head, I mean, sort of from the beginning, with superstitions and taboos and misinterpreted doctrines and laws and your science, your refrigerators, your atomic bombs. But I'm just not buying; you see, I'm just, you just haven't quite succeeded. I think what I'm saying is that, well, I mean, just not conforming, and it's — well, it's just that way.

T: Your feeling at the present time is that you have been very much aware of all the cultural pressures — not always very much aware, but "there have been so many of those in my life — and now I'm going down more deeply into myself to find out what I really feel" and it seems very much at the present time as though that somehow separates you a long ways from your culture, and that's a little frightening, but feels basically good. Is that —

C: Yeah. Well, I have the feeling now that it's okay, really. . . . Then there's something else — a feeling that's starting to grow; well, to be almost formed, as I say. This kind of conclusion, that I'm going to stop looking for something terribly wrong. Now I don't know why. But I mean, just — it's this kind of thing. I'm sort of saying to myself now, well, in view of what I know, what I've found — I'm pretty sure I've ruled out fear, and I'm positive I'm not afraid of shock — I mean, I sort of would have welcomed it. But — in view of the places I've been, what I learned there, then also kind of, well, taking into consideration what I don't know, sort of, maybe this is one of the things that I'll have to date, and say, well, now, I've just — I just can't find it. See? And now without any — without, I should say, any sense of apology or covering up, just sort of simple statement that I can't find what at this time, appears to be bad.

T: Does this catch it? That as you've gone more and more deeply into yourself, and as you think about the kind of things that you've discovered and learned and so on, the conviction grows very, very strong that no matter how far you go, the things that you're going to find are not dire and awful. They have a very different character.

C: Yes, something like that.

Here, even as she recognizes that her feeling goes against the grain of her culture, she feels bound to say that the core of herself is not bad, nor terribly wrong, but something positive. Underneath the layer of controlled surface behavior, underneath the bitterness, underneath the hurt, is a self that is positive, and that is without hate. This I believe is the lesson which our clients have been facing us with for a long time, and which we have been slow to learn.

If hatelessness seems like a rather neutral or negative concept, perhaps we should let Mrs. Oak explain its meaning. In her thirty-ninth interview, as she feels her therapy drawing to a close, she returns to this topic.

C: I wonder if I ought to clarify — it's clear to me, and perhaps that's all that matters really, here, my strong feeling about a hate-free kind of approach. Now that we have brought it up on a rational kind of plane, I know — it sounds negative. And yet in my thinking, my — not really my thinking but my feeling, it — *and* my thinking, yes, my thinking, too — it's a far more positive thing than this — than a love — and it seems to me a far easier kind of a — it's less confining. But it — I realize that it must sort of sound and almost seem like a complete rejection of so many things, of so many creeds and maybe it is. I don't know. But it just to me seems more positive.

T: You can see how it might sound more negative to someone but as far as the meaning that it has for you is concerned, it doesn't seem as binding, as possessive I take it, as love. It seems as though it actually is more — more expandable, more usable, than —

C: Yeah.

T: — any of these narrower terms.

C: Really does to me. It's easier. Well, anyway, it's easier for me to feel that way. And I don't know. It seems to me to really be a way of — of not — of finding yourself in a place where you aren't forced to make rewards and you aren't forced to punish. It is — it means so much. It just seems to me to make for a kind of freedom.

T: M-hm. M-hm. Where one is rid of the need of either rewarding or punishing, then it just seems to you there is so much more freedom for all concerned.

C: That's right. (*Pause*) I'm prepared for some breakdowns along the way.

T: You don't expect it will be smooth sailing.

C: No.

This section is the story — greatly abbreviated — of one client's discovery that the deeper she dug within herself, the less she had to fear; that instead of finding something terribly wrong within herself, she gradually uncovered a core of self which wanted neither to reward nor punish others, a self without hate, a self which was deeply socialized. Do we dare to generalize from this type of experience that if we cut through deeply enough to our organismic nature, that we find that man is a positive and social animal? This is the suggestion from our clinical experience.

Being One's Organism, One's Experience

The thread which runs through much of the foregoing material of this chapter is that psychotherapy (at least client-centered therapy) is a process whereby man becomes his organism — without self-deception, without distortion. What does this mean?

We are talking here about something at an experiential level — a phenomenon which is not easily put into words, and which, if apprehended only at the verbal level, is by that very fact, already distorted. Perhaps if we use several sorts of descriptive formulation, it may ring some bell, however faint, in the reader's experience, and cause him to feel "Oh, now I know, from my own experience, something of what you are talking about."

Therapy seems to mean a getting back to basic sensory and visceral experience. Prior to therapy the person is prone to ask himself, often unwittingly, "What do others think I should do in this situation?" "What would my parents or my culture want me to do?" "What do I think *ought* to be done?" He is thus continually acting in terms of the form which should be imposed upon his behavior. This does not necessarily mean that he always acts in *accord* with the opinions of others. He may indeed endeavor to act so as to contradict the expectations of others. He is nevertheless acting *in terms of* the expectations (often introjected expectations) of others. During the process of therapy the individual comes to ask himself, in regard to ever-widening areas of his life-space, "How do *I* experience this?" "What does it mean to *me*?" "If I behave in a certain way how do I symbolize the meaning which it *will* have for me?" He comes to act on a basis of what may be termed realism — a realistic balancing of the satisfactions and dissatisfactions which any action will bring to himself.

Perhaps it will assist those who, like myself, tend to think in concrete and clinical terms, if I put some of these ideas into schematized formulations of the process through which various clients go. For one client this may mean: "I have thought I must feel only love for my parents, but I find that I experience both love and bitter resentment. Perhaps I can be that person who freely experiences both love *and* resentment." For another client the learning may be: "I have thought I was only bad and worthless. Now I experience myself at times as one of much worth; at other times as one of little worth or usefulness. Perhaps I can be a person who experiences varying degrees of worth." For another: "I have held the conception that no one could really love me for myself. Now I experience the affectional warmth of another for me. Perhaps I can be a person who is lovable by others — perhaps I *am* such a person." For still another: "I have been brought up to feel that I must not appreciate myself — but I do. I can cry for myself, but I can enjoy myself, too. Perhaps I am a richly varied person whom I can enjoy and for whom I can feel sorry." Or, to take the last example from Mrs. Oak, "I have thought that in some deep way I was bad, that the most basic elements in me must be dire and awful. I don't experience that badness, but rather a positive desire to live and let live. Perhaps I can be that person who is, at heart, positive."

What is it that makes possible anything but the first sentence of each of these formulations? It is the addition of awareness. In therapy the person adds to ordinary experience the full and undistorted awareness of his experiencing — of his sensory and visceral reactions. He ceases, or at least decreases, the distortions of experience in awareness. He can be aware of what he is actually experiencing, not simply what he can permit himself to experience after a thorough screening through a conceptual filter. In this sense the person becomes for the first time the full potential of the human organism, with the enriching element of awareness freely added to the basic aspect of sensory and visceral reaction. The person comes to *be* what he *is*, as clients so frequently say in therapy. What this seems to mean is that the individual comes to *be* — in awareness —

what he *is* — in experience. He is, in other words, a complete and fully functioning human organism.

Already I can sense the reactions of some of my readers. "Do you mean that as a result of therapy, man becomes nothing but a human *organism,* a human *animal?* Who will control him? Who will socialize him? Will he then throw over all inhibitions? Have you merely released the beast, the id, in man?" To which the most adequate reply seems to be, "In therapy the individual has actually *become* a human organism, with all the richness which that implies. He is realistically able to control himself, and he is incorrigibly socialized in his desires. There is no beast in man. There is only man in man, and this we have been able to release."

So the basic discovery of psychotherapy seems to me, if our observations have any validity, that we do not need to be afraid of being "merely" homo sapiens. It is the discovery that if we can add to the sensory and visceral experiencing which is characteristic of the whole animal kingdom, the gift of a free and undistorted awareness of which only the human animal seems fully capable, we have an organism which is beautifully and constructively realistic. We have then an organism which is as aware of the demands of the culture as it is of its own physiological demands for food or sex — which is just as aware of its desire for friendly relationships as it is of its desire to aggrandize itself — which is just as aware of its delicate and sensitive tenderness toward others, as it is of its hostilities toward others. When man's unique capacity of awareness is thus functioning freely and fully, we find that we have, not an animal whom we must fear, not a beast who must be controlled, but an organism able to achieve, through the remarkable integrative capacity of its central nervous system, a balanced, realistic, self-enhancing, other-enhancing behavior as a resultant of all these elements of awareness. To put it another way, when man is less than fully man — when he denies to awareness various aspects of his experience — then indeed we have all too often reason to fear him and his behavior, as the present world situation testifies. But when he is most fully man, when he is his complete organism, when awareness of experience, that peculiarly human attribute, is most fully operating, then he is to be trusted, then his behavior is constructive. It is not always conventional. It will not always be conforming. It will be individualized. But it will also be socialized.

A Concluding Comment

I have stated the preceding section as strongly as I am able because it represents a deep conviction growing out of many years of experience. I am quite aware, however, of the difference between conviction and truth. I do not ask anyone to agree with my experience, but only to consider whether the formulation given here agrees with his own experience.

Nor do I apologize for the speculative character of this paper. There is a time for speculation, and a time for the sifting of evidence. It is to be hoped that gradually some of the speculations and opinions and clinical hunches of this paper may be put to operational and definitive test.

References

1. Maslow, A. H. Our maligned animal nature. *Jour. of Psychol.*, 1949, *28*, 273–278.
2. Montagu, A. *On Being Human.* New York: Henry Schuman, Inc., 1950.
3. Rogers, C. R., *Client-Centered Therapy.* Boston: Houghton Mifflin Co., 1951, Chapter IV, "The Process of Therapy."

6

What It Means to Become a Person

This chapter was first given as a talk to a meeting at Oberlin College in 1954. I was trying to pull together in more completely organized form, some of the conceptions of therapy which had been growing in me. I have revised it slightly.

As is customary with me, I was trying to keep my thinking close to the grass roots of actual experience in therapeutic interviews, so I drew heavily upon recorded interviews as the source of the generalizations which I make.

IN MY WORK at the Counseling Center of the University of Chicago, I have the opportunity of working with people who present a wide variety of personal problems. There is the student concerned about failing in college; the housewife disturbed about her marriage; the individual who feels he is teetering on the edge of a complete breakdown or psychosis; the responsible professional man who spends much of his time in sexual fantasies and functions inefficiently in his work; the brilliant student, at the top of his class, who is paralyzed by the conviction that he is hopelessly and helplessly inadequate; the parent who is distressed by his child's behavior; the popular girl who finds herself unaccountably overtaken by sharp spells of black depression; the woman who fears that life and love are passing her by, and that her good graduate record is a poor recompense; the man who has become convinced that powerful or sinister forces are plotting against him; — I could go on and on with the many different and unique problems which people bring to us. They run the gamut of life's experiences. Yet there is no satisfaction in giving this type of catalog, for, as counselor, I know that the problem as stated in the first interview will not be the problem as seen in the second or third hour, and by the tenth interview it will be a still different problem or series of problems.

I have however come to believe that in spite of this bewildering horizontal multiplicity, and the layer upon layer of vertical complexity, there is perhaps only one problem. As I follow the experience of many clients in the therapeutic relationship which we endeavor to create for them, it seems to me that each one is raising the same question. Below the level of the problem situation about which the individual is complaining — behind the trouble with studies, or wife, or employer, or with his own uncontrollable or bizarre behavior, or with his frightening feelings, lies one central search. It seems to me that at bottom each person is asking, "Who am I, *really?* How can I get in touch with this real self, underlying all my surface behavior? How can I become myself?"

THE PROCESS OF BECOMING

GETTING BEHIND THE MASK

Let me try to explain what I mean when I say that it appears that the goal the individual most wishes to achieve, the end which he knowingly and unknowingly pursues, is to become himself.

When a person comes to me, troubled by his unique combination of difficulties, I have found it most worth while to try to create a relationship with him in which he is safe and free. It is my purpose to understand the way he feels in his own inner world, to accept him as he is, to create an atmosphere of freedom in which he can move in his thinking and feeling and being, in any direction he desires. How does he use this freedom?

It is my experience that he uses it to become more and more himself. He begins to drop the false fronts, or the masks, or the roles, with which he has faced life. He appears to be trying to discover something more basic, something more truly himself. At first he lays aside masks which he is to some degree aware of using. One young woman student describes in a counseling interview one of the masks she has been using, and how uncertain she is whether underneath this appeasing, ingratiating front there is any real self with convictions.

> I was thinking about this business of standards. I somehow developed a sort of knack, I guess, of — well — habit — of trying to make people feel at ease around me, or to make things go along smoothly. There always had to be some appeaser around, being sorta the oil that soothed the waters. At a small meeting, or a little party, or something — I could help things go along nicely and appear to be having a good time. And sometimes I'd surprise myself by arguing against what I really thought when I saw that the person in charge would be quite unhappy about it if I didn't. In other words I just wasn't ever — I mean, I didn't find myself ever being set and definite about things. Now the reason why I did it probably was I'd been doing it around home so much. I just didn't stand up for my own convictions, until I don't know whether I have any convictions to stand up for. I haven't been really honestly being myself, or actually knowing what my real self is, and I've been just playing a sort of false role.

You can, in this excerpt, see her examining the mask she has been using, recognizing her dissatisfaction with it, and wondering how to get to the real self underneath, if such a self exists.

In this attempt to discover his own self, the client typically uses the relationship to explore, to examine the various aspects of his own experience, to recognize and face up to the deep contradictions which he often discovers. He learns how much of his behavior, even how much of the feeling he experiences, is not real, is not something which flows from the genuine reactions of his organism, but is a façade, a front, behind which he has been hiding. He discovers how much of his life is guided by what he thinks he *should* be, not by what he is. Often he discovers that he exists only in response to the demands of others, that he seems to have no self of his own, that he is only trying to think, and feel, and behave in the way that others believe he *ought* to think, and feel and behave.

In this connection I have been astonished to find how accurately the Danish philosopher, Søren Kierkegaard, pictured the dilemma of the individual more than a century ago, with keen psychological insight. He points out that the most common despair is to be in despair at not choosing, or willing, to be oneself; but that the deepest form of despair is to choose "to be another than himself." On the other hand "to will to be that self which one truly is, is indeed the opposite of despair," and this choice is the deepest responsibility of man. As I read some of his writings I almost feel that he must have listened in on the statements made by our clients as they search and explore for the reality of self — often a painful and troubling search.

This exploration becomes even more disturbing when they find themselves involved in removing the false faces which they had not known were false faces. They begin to engage in the frightening task of exploring the turbulent and sometimes violent feelings within themselves. To remove a mask which you had thought was part of your real self can be a deeply disturbing experience, yet when there is freedom to think and feel and be, the individual moves toward such a goal. A few statements from a person who had completed a series of psychotherapeutic interviews, will illustrate this. She uses many metaphors as she tells how she struggled to get to the core of herself.

> As I look at it now, I was peeling off layer after layer of defenses. I'd build them up, try them, and then discard them when you remained the same. I didn't know what was at the bottom and I was very much afraid to find out, but I *had* to keep on trying. At first I felt there was nothing within me — just a great emptiness where I needed and wanted a solid core. Then I began to feel that

> I was facing a solid brick wall, too high to get over and too thick to go through. One day the wall became translucent, rather than solid. After this, the wall seemed to disappear but beyond it I discovered a dam holding back violent, churning waters. I felt as if I were holding back the force of these waters and if I opened even a tiny hole I and all about me would be destroyed in the ensuing torrent of feelings represented by the water. Finally I could stand the strain no longer and I let go. All I did, actually, was to succumb to complete and utter self pity, then hate, then love. After this experience, I felt as if I had leaped a brink and was safely on the other side, though still tottering a bit on the edge. I don't know what I was searching for or where I was going, but I felt then as I have always felt whenever I really lived, that I was moving forward.

I believe this represents rather well the feelings of many an individual that if the false front, the wall, the dam, is not maintained, then everything will be swept away in the violence of the feelings that he discovers pent-up in his private world. Yet it also illustrates the compelling necessity which the individual feels to search for and become himself. It also begins to indicate the way in which the individual determines the reality in himself — that when he fully experiences the feelings which at an organic level he *is*, as this client experienced her self-pity, hatred, and love, then he feels an assurance that he is being a part of his real self.

The Experiencing of Feeling

I would like to say something more about this experiencing of feeling. It is really the discovery of unknown elements of self. The phenomenon I am trying to describe is something which I think is quite difficult to get across in any meaningful way. In our daily lives there are a thousand and one reasons for not letting ourselves experience our attitudes fully, reasons from our past and from the present, reasons that reside within the social situation. It seems too dangerous, too potentially damaging, to experience them freely and fully. But in the safety and freedom of the therapeutic relationship, they can be experienced fully, clear to the limit of what they are. They can be and are experienced in a fashion that I like to think of as a "pure culture," so that for the moment the person *is* his fear, or he *is* his anger, or he *is* his tenderness, or whatever.

Perhaps again I can clarify this by giving an example from a client which will indicate and convey something of what I mean. A young man, a graduate student who is deep in therapy, has been puzzling over a vague feeling which he senses in himself. He gradually identifies it as a frightened feeling of some kind, a fear of failing, a fear of not getting his Ph.D. Then comes a long pause. From this point on we will let the recorded interview speak for itself.

Client: I was kinda letting it seep through. But I also tied it in with you and with my relationship with you. And that's one thing I feel about it is kind of a fear of it going away; or that's another thing — it's so hard to get hold of — there's kind of two pulling feelings about it. Or two "me's" somehow. One is the scared one that wants to hold on to things, and that one I guess I can feel pretty clearly right now. You know, I kinda need things to hold on to — and I feel kinda scared.

Therapist: M-hm. That's something you can feel right this minute, and have been feeling and perhaps *are* feeling in regard to our relationship, too.

C: Won't you let me *have* this, because, you know, I kinda *need* it. I can be so lonely and scared without it.

T: M-hm, m-hm. Let me hang on to this because I'd be terribly scared if I didn't. Let me *hold* on to it. (*Pause*)

C: It's kinda the same thing — *Won't* you let me have my thesis or my Ph.D. so then . . . 'Cause I kinda *need* that little world. I mean. . . .

T: In both instances it's kind of a pleading thing too, isn't it? Let me *have* this because I need it *badly*. I'd be awfully frightened without it. (*Long pause.*)

C: I get a sense of . . . I can't somehow get much further . . . It's this kind of *pleading* little boy, somehow, even . . . What's this gesture of begging? (*Putting his hands together as if in prayer*) Isn't it funny? 'Cause that . . .

T: You put your hands in sort of a supplication.

C: Ya, that's right! Won't you *do* this for me, kinda . . . Oh, that's *terrible!* Who, me, *beg?*

Perhaps this excerpt will convey a bit of the thing I have been talking about, the experiencing of a feeling all the way to the limit. Here he is, for a moment, experiencing himself as nothing but a pleading little boy, supplicating, begging, dependent. At that moment he is nothing but his pleadingness, all the way through. To be sure he almost immediately backs away from this experiencing by saying "Who, me, *beg?*" but it has left its mark. As he says a moment later, "It's such a wondrous thing to have these new things come out of me. It amazes me so much each time, and then again there's that same feeling, kind of feeling scared that I've so much of this that I'm keeping back or something." He realizes that this has bubbled through, and that for the moment he *is* his dependency, in a way which astonishes him.

It is not only dependency that is experienced in this all-out kind of fashion. It may be hurt, or sorrow, or jealousy, or destructive anger, or deep desire, or confidence and pride, or sensitive tenderness, or outgoing love. It may be any of the emotions of which man is capable.

What I have gradually learned from experiences such as this, is that the individual in such a moment, is coming to *be* what he *is*. When a person has, throughout therapy, experienced in this fashion all the emotions which organismically arise in him, and has experienced them in this knowing and open manner, then he has experienced *himself*, in all the richness that exists within himself. He has become what he is.

The Discovery of Self in Experience

Let us pursue a bit further this question of what it means to become one's self. It is a most perplexing question and again I will try to take from a statement by a client, written between interviews, a suggestion of an answer. She tells how the various façades by which she has been living have somehow crumpled and collapsed, bringing a feeling of confusion, but also a feeling of relief. She continues:

> You know, it seems as if all the energy that went into holding the arbitrary pattern together was quite unnecessary — a waste. You think you have to make the pattern yourself; but there are so many pieces, and it's so hard to see where they fit. Sometimes you put them in the wrong place, and the more pieces mis-fitted, the more effort it takes to hold them in place, until at last you are so tired that even that awful confusion is better than holding on any longer. Then you discover that left to themselves the jumbled pieces fall quite naturally into their own places, and a living pattern emerges without any effort at all on your part. Your job is just to discover it, and in the course of that, you will find yourself and your own place. You must even let your own experience tell you its own meaning; the minute *you* tell it what it means, you are at war with yourself.

Let me see if I can take her poetic expression and translate it into the meaning it has for me. I believe she is saying that to be herself means to find the pattern, the underlying order, which exists in the ceaselessly changing flow of her experience. Rather than to try to hold her experience into the form of a mask, or to make it be a form or structure that it is not, being herself means to discover the unity and harmony which exists in her own actual feelings and reactions. It means that the real self is something which is comfortably discovered in one's experiences, not something imposed upon it.

Through giving excerpts from the statements of these clients, I have been trying to suggest what happens in the warmth and understanding of a facilitating relationship with a therapist. It seems that gradually, painfully, the individual explores what is behind the masks he presents to the world, and even behind the masks with which he has been deceiving himself. Deeply and often vividly he experiences the various elements of himself which have been hidden within. Thus to an increasing degree he becomes himself — not a façade of conformity to others, not a cynical denial of all feeling, nor a front of intellectual rationality, but a living, breathing, feeling, fluctuating process — in short, he becomes a person.

The Person Who Emerges

I imagine that some of you are asking, "But what *kind* of a person does he become? It isn't enough to say that he drops the façades. What kind of person lies underneath?" Since one of the most obvious facts is that each individual tends to become a separate and distinct and unique person, the answer is not easy. However I would like to point out some of the characteristic trends which I see. No one person would fully exemplify these characteristics, no one person fully achives the description I will give, but I do see certain generalizations which can be drawn, based upon living a therapeutic relationship with many clients.

Openness to Experience

First of all I would say that in this process the individual becomes more open to his experience. This is a phrase which has come to have a great deal of meaning to me. It is the opposite of defensiveness. Psychological research has shown that if the evidence of our senses runs contrary to our picture of self, then that evidence is distorted. In other words we cannot see all that our senses report, but only the things which fit the picture we have.

Now in a safe relationship of the sort I have described, this defensiveness or rigidity, tends to be replaced by an increasing openness to experience. The individual becomes more openly aware of his own feelings and attitudes as they exist in him at an organic level, in the way I tried to describe. He also becomes more aware of reality as it exists outside of himself, instead of perceiving it in preconceived categories. He sees that not all trees are green, not all men are stern fathers, not all women are rejecting, not all failure experiences prove that he is no good, and the like. He is able to take in the evidence in a new situation, *as it is*, rather than distorting it to fit a pattern which he already holds. As you might expect, this increasing ability to be open to experience makes him far more realistic in dealing with new people, new situations, new problems. It means that his beliefs are not rigid, that he can tolerate ambiguity. He can receive much conflicting evidence without forcing closure upon the situation. This openness of awareness to what exists at *this moment* in *oneself* and in *the situation* is, I believe, an important element in the description of the person who emerges from therapy.

Perhaps I can give this concept a more vivid meaning if I illustrate it from a recorded interview. A young professional man reports in the 48th interview the way in which he has become more open to some of his bodily sensations, as well as other feelings.

C: It doesn't seem to me that it would be possible for anybody to relate all the changes that you feel. But I certainly have felt recently that I have more respect for, more objectivity toward my physical makeup. I mean I don't expect too much of myself. This is how it works out: It feels to me that in the past I used to fight a certain tiredness that I felt after supper. Well, now I feel pretty sure that I really *am tired* — that I am not making myself tired — that I am just physiologically lower. It seemed that I was just constantly criticizing my tiredness.

T: So you can let yourself *be* tired, instead of feeling along with it a kind of criticism of it.

C: Yes, that I shouldn't be tired or something. And it seems in a way to be pretty profound that I can just not fight this tiredness, and along with it goes a real feeling of *I've* got to slow down, too, so that being tired isn't such an awful thing. I think I can also kind of pick up a thread here of why I should be that way in the way my father is and the way he looks at some of these things. For instance, say that I was sick, and I would report this, and it would seem that overtly he would want to do something about it but he would also communicate, "Oh, my gosh, more trouble." You know, something like that.

T: As though there were something quite annoying really about being physically ill.

C: Yeah, I'm sure that my father has the same disrespect for his own physiology that I have had. Now last summer I twisted my back, I wrenched it, I heard it snap and everything. There was real pain there all the time at first, real sharp. And I had the doctor look at it and he said it wasn't serious, it should heal by itself as long as I didn't bend too much. Well this was months ago — and I have been noticing recently that — hell, this is a real pain and it's still there — and it's not my fault.

T: It doesn't prove something bad about you —

C: No — and one of the reasons I seem to get more tired than I should maybe is because of this constant strain, and so — I have already made an appointment with one of the doctors at the hospital that he would look at it and take an X ray or something. In a way I guesss you could say that I am just more accurately sensitive — or objectively sensitive to this kind of thing. . . . And this is really a profound change as I say, and of course my relationship with my wife and the two children is — well, you just wouldn't recognize it if you could see me inside — as you have — I mean — there just doesn't seem to be anything more wonderful than really and genuinely — really *feeling* love for your own children and at the same time receiving it. I don't know how to put this. We have such an increased respect — both of us — for Judy and we've noticed just — as we participated in this — we have noticed such a tremendous change in her — it seems to be a pretty deep kind of thing.

T: It seems to me you are saying that you can listen more accurately to yourself. If your body says it's tired, you listen to it and believe it, instead of criticizing it; if it's in pain, you can listen to that; if the feeling is really loving your wife or children, you can *feel* that, and it seems to show up in the differences in them too.

Here, in a relatively minor but symbolically important excerpt, can be seen much of what I have been trying to say about openness to experience. Formerly he could not freely feel pain or illness, because being ill meant being unacceptable. Neither could he feel tenderness and love for his child, because such feelings meant being weak, and he had to maintain his façade of being strong. But now he can be genuinely open to the experiences of his organism — he can be tired when he is tired, he can feel pain when his organism is in pain, he can freely experience the love he feels for his daughter, and he can also feel and express annoyance toward her, as he goes on to say in the next portion of the interview. He can fully live the experiences of his total organism, rather than shutting them out of awareness.

TRUST IN ONE'S ORGANISM

A second characteristic of the persons who emerge from therapy is difficult to describe. It seems that the person increasingly discovers that his own organism is trustworthy, that it is a suitable instrument for discovering the most satisfying behavior in each immediate situation.

If this seems strange, let me try to state it more fully. Perhaps it will help to understand my description if you think of the individual as faced with some existential choice: "Shall I go home to my family during vacation, or strike out on my own?" "Shall I drink this third cocktail which is being offered?" "Is this the person whom I would like to have as my partner in love and in life?" Thinking of such situations, what seems to be true of the person who emerges from the therapeutic process? To the extent that this person is open to all of his experience, he has access to all of the available data in the situation, on which to base his behavior. He has knowledge of his own feelings and impulses, which are often complex and contradictory. He is freely able to sense the social demands, from the relatively rigid social "laws" to the desires of friends and family. He has access to his memories of similar situations, and the consequences of different behaviors in those situations. He has a relatively accurate perception of this external situation in all of its complexity. He is better able to permit his total organism, his conscious thought participating, to consider, weigh and balance each stimulus, need, and demand, and its relative weight and intensity. Out of this complex weighing and balancing he is able to discover that course of action which seems to come closest to satisfying all his needs in the situation, long-range as well as immediate needs.

In such a weighing and balancing of all of the components of a given life choice, his organism would not by any means be infallible. Mistaken choices might be made. But because he tends to be open to his experience, there is a greater and more immediate awareness of unsatisfying consequences, a quicker correction of choices which are in error.

It may help to realize that in most of us the defects which interfere with this weighing and balancing are that we include things that are not a part of our experience, and exclude elements which are. Thus an individual may persist in the concept that "I can handle liquor," when openness to his past experience would indicate that this is scarcely correct. Or a young woman may see only the good qualities of her prospective mate, where an openness to experience would indicate that he possesses faults as well.

In general then, it appears to be true that when a client is open to his experience, he comes to find his organism more trustworthy. He feels less fear of the emotional reactions which he has. There is a gradual growth of trust in, and even affection for the complex, rich, varied assortment of feelings and tendencies which exist in him at the organic level. Consciousness, instead of being the watchman over a dangerous and unpredictable lot of impulses, of which few can be permitted to see the light of day, becomes the comfortable inhabitant of a society of impulses and feelings and thoughts, which are discovered to be very satisfactorily self-governing when not fearfully guarded.

An Internal Locus of Evaluation

Another trend which is evident in this process of becoming a person relates to the source or locus of choices and decisions, or evaluative judgments. The individual increasingly comes to feel that this locus of evaluation lies within himself. Less and less does he look to others for approval or disapproval; for standards to live by; for decisions and choices. He recognizes that it rests within himself to choose; that the only question which matters is, "Am I living in a way which is deeply satisfying to me, and which truly expresses me?" This I think is perhaps *the* most important question for the creative individual.

Perhaps it will help if I give an illustration. I would like to give a brief portion of a recorded interview with a young woman, a graduate student, who had come for counseling help. She was initially very much disturbed about many problems, and had been contemplating suicide. During the interview one of the feelings she discovered was her great desire to be dependent, just to let someone else take over the direction of her life. She was very critical of those who had not given her enough guidance. She talked about one after another of her professors, feeling bitterly that none of them had taught her anything with deep meaning. Gradually she began to realize that part of the difficulty was the fact that she had taken no initiative in *participating* in these classes. Then comes the portion I wish to quote.

I think you will find that this excerpt gives you some indication of what it means in experience to accept the locus of evaluation as being within oneself. Here then is the quotation from one of the later interviews with this young woman as she has begun to realize that perhaps she is partly responsible for the deficiencies in her own education.

C: Well now, I wonder if I've been going around doing that, getting smatterings of things, and not getting hold, not really getting down to things.

T: Maybe you've been getting just spoonfuls here and there rather than really digging in somewhere rather deeply.

C: M-hm. That's why I say — (*slowly and very thoughtfully*) well, with that sort of a foundation, well, it's really up to *me.* I mean, it seems to be really apparent to me that I *can't depend on someone else* to give me an education. (*Very softly*) I'll really have to get it myself.

T: It really begins to come home — there's only one person that can educate you — a realization that perhaps nobody else *can give* you an education.

C: M-hm. (*Long pause — while she sits thinking*) I have all the symptoms of fright. (*Laughs softly*)

T: Fright? That this is a scary thing, is that what you mean?

C: M-hm. (*Very long pause — obviously struggling with feelings in herself*).

T: Do you want to say any more about what you mean by that? That it really does give you the symptoms of fright?

C: (*Laughs*) I, uh — I don't know whether I quite know. I mean — well it really seems like I'm cut loose (*pause*), and it seems that I'm very — I don't know — in a vulnerable position, but I, uh, I brought this up and it, uh, somehow it almost came out without my saying it. It seems to be — it's something I let out.

T: Hardly a part of you.

C: Well, I felt surprised.

T: As though, "Well for goodness sake, did I say that?" (*Both chuckle.*)

C: Really, I don't think I've had that feeling before. I've — uh, well, this really feels like I'm saying something that, uh, *is* a part of me really. (*Pause*) Or, uh, (*quite perplexed*) it feels like I sort of have, uh, I don't know. I have a feeling of *strength*, and yet, I have a feeling of — realizing it's so sort of fearful, of fright.

T: That is, do you mean that saying something of that sort gives you at the same time a feeling of, of strength in saying it, and yet at the same time a frightened feeling of *what* you have said, is that it?

C: M-hm. I am feeling that. For instance, I'm feeling it internally now — a sort of surging up, or force or outlet. As if that's something really big and strong. And yet, uh, well at first it was almost a physical feeling of just being out alone, and sort of cut off from a — a support I had been carrying around.

T: You feel that it's something deep and strong, and surging forth, and at the same time, you just feel as though you'd cut yourself loose from any support when you say it.

C: M-hm. Maybe that's — I don't know — it's a disturbance of a kind of pattern I've been carrying around, I think.

T: It sort of shakes a rather significant pattern, jars it loose.

C: M-hm. (*Pause, then cautiously, but with conviction*) I, I think — I don't know, but I have the feeling that then I am going to begin to *do* more things that I know I should do. . . . There are so many things that I need to do. It seems in so many avenues of my living I have to work out new ways of behavior, but — maybe — I can see myself doing a little better in some things

I hope that this illustration gives some sense of the strength which is experienced in being a unique person, responsible for oneself, and also the uneasiness that accompanies this assumption of responsibility. To recognize that "I am the one who chooses" and "I am the one who determines the value of an experience for me" is both an invigorating and a frightening realization.

WILLINGNESS TO BE A PROCESS

I should like to point out one final characteristic of these individuals as they strive to discover and become themselves. It is that the individual seems to become more content to be a *process* rather than a *product.* When he enters the therapeutic relationship, the client is likely to wish to achieve some fixed state: he wants to reach the point where his problems are solved, or where he is effective in his work, or where his marriage is satisfactory. He tends, in the freedom of the therapeutic relationship to drop such fixed goals, and to accept a more satisfying realization that he is not a fixed entity, but a process of becoming.

One client, at the conclusion of therapy, says in rather puzzled fashion, "I haven't finished the job of integrating and reorganizing myself, but that's only confusing, not discouraging, now that I realize this is a continuing process. . . . It's exciting, sometimes upsetting, but deeply encouraging to feel yourself in action, apparently knowing where you are going even though you don't always consciously know where that is." One can see here both the expression of trust in the organism, which I have mentioned, and also the realization of self as a process. Here is a personal description of what it seems like to accept oneself as a stream of becoming, not a finished product. It means that a person is a fluid process, not a fixed and static entity; a flowing river of change, not a block of solid material; a continually changing constellation of potentialities, not a fixed quantity of traits.

Here is another statement of this same element of fluidity or existential living, "This whole train of experiencing, and the meanings that I have thus far discovered in it, seem to have launched me on a process which is both fascinating and at times a little frightening. It seems to mean letting my experiences carry me on, in a direction which appears to be forward, towards goals that I can but dimly define, as I try to understand at least the current meaning of that experience. The sensation is that of floating with a complex stream of experience, with the fascinating possibility of trying to comprehend its ever-changing complexity."

Conclusion

I have tried to tell you what has seemed to occur in the lives of people with whom I have had the privilege of being in a relationship as they struggled toward becoming themselves. I have endeavored to describe, as accurately as I can, the meanings which seem to be involved in this process of becoming a person. I am sure that this process is *not* one that occurs only in therapy. I am sure that I do not see it clearly or completely, since I keep changing my comprehension and understanding of it. I hope you will accept it as a current and tentative picture, not as something final.

One reason for stressing the tentative nature of what I have said is that I wish to make it clear that I am *not* saying, "This is what you should become; here is the goal for you." Rather, I am saying that these are some of the meanings I see in the experiences that my clients and I have shared. Perhaps this picture of the experience of others may illuminate or give more meaning to some of your own experience.

I have pointed out that each individual appears to be asking a double question: "Who am I?" and "How may I become myself?" I have stated that in a favorable psychological climate a process of becoming takes place; that here the individual drops one after another of the defensive masks with which he has faced life; that he experiences fully the hidden aspects of himself; that he discovers in these experiences the stranger who has been living behind these masks, the stranger who is himself. I have tried to give my picture of the characteristic attributes of the person who emerges; a person who is more open to all of the elements of his organic experience; a person who is developing a trust in his own organism as an instrument of sensitive living; a person who accepts the locus of evaluation as residing within himself; a person who is learning to live in his life as a participant in a fluid, ongoing process, in which he is continually discovering new aspects of himself in the flow of his experience. These are some of the elements which seem to me to be involved in becoming a person.

7

A Process Conception of Psychotherapy

In the autumn of 1956 I was greatly honored by the American Psychological Association, which bestowed upon me one of its first three Distinguished Scientific Contribution Awards. There was however a penalty attached to the award, which was that one year later, each recipient was to present a paper to the Association. It did not appeal to me to review work which we had done in the past. I decided rather to devote the year to a fresh attempt to understand the process by which personality changes. I did this, but as the next autumn approached, I realized that the ideas I had formed were still unclear, tentative, hardly in shape for presentation. Nevertheless I tried to set down the jumbled sensings which had been important to me, out of which was emerging a concept of process different from anything I had clearly perceived before. When I had finished I found I had a paper much too long to deliver, so I cut it down to an abbreviated form for presentation on September 2, 1957 to the American Psychological Convention in New York. The present chapter is neither as long as the initial form, nor as abbreviated as the second form.

It will be discovered that though the two preceding chapters view the process of therapy almost entirely from a phenomenological point of view, from within the client's frame of reference, this for-

mulation endeavors to capture those qualities of expression which may be observed by another, and hence views it more from an external frame of reference.

Out of the observations recorded in this paper a "Scale of Process in Psychotherapy" has been developed which can be applied operationally to excerpts from recorded interviews. It is still in process of revision and improvement. Even in its present form it has reasonable inter-judge reliability, and gives meaningful results. Cases which by other criteria are known to be more successful, show greater movement on the Process Scale than less successful cases. Also, to our surprise it has been found that successful cases begin *at a higher level on the Process Scale than do unsuccessful cases. Evidently we do not yet know, with any satisfactory degree of assurance, how to be of therapeutic help to individuals whose behavior when they come to us is typical of stages one and two as described in this chapter. Thus the ideas of this paper, poorly formed and incomplete as they seemed to me at the time, are already opening up new and challenging areas for thought and investigation.*

THE PUZZLE OF PROCESS

I WOULD LIKE to take you with me on a journey of exploration. The object of the trip, the goal of the search, is to try to learn something of the *process* of psychotherapy, or the *process* by which personality change takes place. I would warn you that the goal has not yet been achieved, and that it seems as though the expedition has advanced only a few short miles into the jungle. Yet perhaps if I can take you with me, you will be tempted to discover new and profitable avenues of further advance.

My own reason for engaging in such a search seems simple to me. Just as many psychologists have been interested in the invariant aspects of personality — the unchanging aspects of intelligence, temperament, personality structure — so I have long been interested in the invariant aspects of *change* in personality. Do personality and behavior change? What commonalities exist in such changes? What commonalities exist in the conditions which precede change? Most important of all, what is the process by which such change occurs?

Until recently we have for the most part tried to learn something of this process by studying outcomes. We have many facts, for example, regarding the changes which take place in self-perception, or in perception of others. We have not only measured these changes over the whole course of therapy, but at intervals during therapy. Yet even this last gives us little clue as to the *process* involved. Studies of segmented outcomes are still measures of outcome, giving little knowledge of the way in which the change takes place.

Puzzling over this problem of getting at the process has led me to realize how little objective research deals with process in any field. Objective research slices through the frozen moment to provide us with an exact picture of the inter-relationships which exist at that moment. But our understanding of the ongoing movement — whether it be the process of fermentation, or the circulation of the blood, or the process of atomic fission — is generally provided by a theoretical formulation, often supplemented, where feasible, with a clinical observation of the process. I have thus come to realize that perhaps I am hoping for too much to expect that research procedures can shed light directly upon the process of personality change. Perhaps only theory can do that.

A Rejected Method

When I determined, more than a year ago, to make a fresh attempt to understand the way in which such change takes place, I first considered various ways in which the experience of therapy might be described in terms of some other theoretical framework. There was much that was appealing in the field of communication theory, with its concepts of feedback, input and output signals, and the like. There was the possibility of describing the process of therapy in terms of learning theory, or in terms of general systems theory. As I studied these avenues of understanding I became convinced that it would be possible to translate the process of psychotherapy into any one of these theoretical frameworks. It would, I believe, have certain advantages to do so. But I also became convinced that in a field so new, this is not what is most needed.

I came to a conclusion which others have reached before, that in a new field perhaps what is needed first is to steep oneself in the *events*, to approach the phenomena with as few preconceptions as possible, to take a naturalist's observational, descriptive approach to these events, and to draw forth those low-level inferences which seem most native to the material itself.

THE MODE OF APPROACH

So, for the past year, I have used the method which so many of us use for generating hypotheses, a method which psychologists in this country seem so reluctant to expose or comment on. I used myself as a tool.

As a tool, I have qualities both good and bad. For many years I have experienced therapy as a therapist. I have experienced it on the other side of the desk as a client. I have thought about therapy, carried on research in this field, been intimately acquainted with the research of others. But I have also formed biases, have come to have a particular slant on therapy, have tried to develop theoretical abstractions regarding therapy. These views and theories would tend to make me less sensitive to the events themselves. Could I open myself to the phenomena of therapy freshly, naively? Could I let the totality of my experience be as effective a tool as it might potentially be, or would my biases prevent me from seeing what was there? I could only go ahead and make the attempt.

So, during this past year I have spent many hours listening to recorded therapeutic interviews — trying to listen as naively as possible. I have endeavored to soak up all the clues I could capture as to the process, as to what elements are significant in change. Then I have tried to abstract from that sensing the simplest abstractions which would describe them. Here I have been much stimulated and helped by the thinking of many of my colleagues, but I would like to mention my special indebtedness to Eugene Gendlin, William Kirtner and Fred Zimring, whose demonstrated ability to think in new ways about these matters has been particularly helpful, and from whom I have borrowed heavily.

The next step has been to take these observations and low-level abstractions and formulate them in such a way that testable hypotheses can readily be drawn from them. This is the point I have reached. I make no apology for the fact that I am reporting no empirical investigations of these formulations. If past experience is any guide, then I may rest assured that, if the formulations I am about to present check in any way with the subjective experience of other therapists, then a great deal of research will be stimulated, and in a few years there will be ample evidence of the degree of truth and falsity in the statements which follow.

The Difficulties and Excitement of the Search

It may seem strange to you that I tell you so much of the personal process I went through in seeking for some simple — and I am sure, inadequate — formulations. It is because I feel that nine-tenths of research is always submerged, and that only the iciest portion is ever seen, a very misleading segment. Only occasionally does someone like Mooney (6, 7) describe the whole of the research method as it exists in the individual. I too should like to reveal something of the whole of this study as it went on in me, not simply the impersonal portion.

Indeed I wish I might share with you much more fully some of the excitement and discouragement of this effort to understand process. I would like to tell you of my fresh discovery of the way feelings "hit" clients — a word they frequently use. The client is talking about something of importance, when wham! he is hit by a feeling — not something named or labelled but an experiencing of an unknown something which has to be cautiously explored before it can be named at all. As one client says, "It's a feeling that I'm caught with. I can't even know what it connects with." The frequency of this event was striking to me.

Another matter of interest was the variety of ways in which clients do come closer to their feelings. Feelings "bubble up through," they "seep through." The client also lets himself "down into" his feeling, often with caution and fear. "I want to get down into this feeling. You can kinda see how hard it is to get really close to it."

Still another of these naturalistic observations has to do with the importance which the client comes to attach to *exactness* of symbolization. He wants just the precise word which for him describes the feeling he has experienced. An approximation will not do. And this is certainly for clearer communication within himself, since any one of several words would convey the meaning equally well to another.

I came also to appreciate what I think of as "moments of movement" — moments when it appears that change actually occurs. These moments, with their rather obvious physiological concomitants, I will try to describe later.

I would also like to mention the profound sense of despair I sometimes felt, wandering naively in the incredible complexity of the therapeutic relationship. Small wonder that we prefer to approach therapy with many rigid preconceptions. We feel we must bring order *to* it. We can scarcely dare to hope that we can discover order *in* it.

These are a few of the personal discoveries, puzzlements, and discouragements which I encountered in working on this problem. Out of these came the more formal ideas which I would now like to present.

A Basic Condition

If we were studying the process of growth in plants, we would assume certain constant conditions of temperature, moisture and sunlight, in forming our conceptualization of the process. Likewise in conceptualizing the process of personality change in psychotherapy, I shall assume a constant and optimal set of conditions for facilitating this change. I have recently tried to spell out these conditions in some detail (8). For our present purpose I believe I can state this assumed condition in one word. Throughout the discussion which follows, I shall assume that the client experiences himself as being fully *received*. By this I mean that whatever his feelings — fear, despair, insecurity, anger, whatever his mode of expression — silence, gestures, tears, or words; whatever he finds himself being in this moment, he senses that he is psychologically *received*, just as he is, by the therapist. There is implied in this term the concept of being understood, empathically, and the concept of acceptance. It is also well to point out that it is the client's experience of this condition which makes it optimal, not merely the fact of its existence in the therapist.

In all that I shall say, then, about the process of change, I shall assume as a constant an optimal and maximum condition of being received.

The Emerging Continuum

In trying to grasp and conceptualize the process of change, I was initially looking for elements which would mark or characterize change itself. I was thinking of change as an entity, and searching for its specific attributes. What gradually emerged in my understanding as I exposed myself to the raw material of change was a continuum of a different sort than I had conceptualized before.

Individuals move, I began to see, not from a fixity or homeostasis through change to a new fixity, though such a process is indeed possible. But much the more significant continuum is from fixity to changingness, from rigid structure to flow, from stasis to process. I formed the tentative hypothesis that perhaps the qualities of the client's expression at any one point might indicate his position on this continuum, might indicate where he stood in the process of change.

I gradually developed this concept of a process, discriminating seven stages in it, though I would stress that it is a continuum, and that whether one discriminated three stages or fifty, there would still be all the intermediate points.

I came to feel that a given client, taken as a whole, usually exhibits behaviors which cluster about a relatively narrow range on this continuum. That is, it is unlikely that in one area of his life the client would exhibit complete fixity, and in another area complete changingness. He would tend, as a whole, to be at some stage in this process. However, the process I wish to describe applies more exactly, I believe, to given areas of personal meanings, where I hypothesize that the client would, in such an area, be quite definitely at one stage, and would not exhibit characteristics of various stages.

SEVEN STAGES OF PROCESS

Let me then try to portray the way in which I see the successive stages of the process by which the individual changes from fixity to flowingness, from a point nearer the rigid end of the continuum to a point nearer the "in-motion" end of the continuum. If I am correct in my observations then it is possible that by dipping in and sampling the qualities of experiencing and expressing in a given individual, in a climate where he feels himself to be completely received, we may be able to determine where he is in this continuum of personality change.

FIRST STAGE

The individual in this stage of fixity and remoteness of experiencing is not likely to come voluntarily for therapy. However I can to some degree illustrate the characteristics of this stage.

There is an unwillingness to communicate self. Communication is only about externals.

Example: "Well, I'll tell you, it always seems a little bit nonsensical to talk about one's self except in times of dire necessity."*

Feelings and personal meanings are neither recognized nor owned.
Personal constructs (to borrow Kelly's helpful term (3)) are extremely rigid.
Close and communicative relationships are construed as dangerous.
No problems are recognized or perceived at this stage.
There is no desire to change.

Example: "I think I'm practically healthy."

There is much blockage of internal communication.

Perhaps these brief statements and examples will convey something of the psychological fixity of this end of the continuum. The

* The many examples used as illustrations are taken from recorded interviews, unless otherwise noted. For the most part they are taken from interviews which have never been published, but a number of them are taken from the report of two cases by Lewis, Rogers and Shlien (5).

individual has little or no recognition of the ebb and flow of the feeling life within him. The ways in which he construes experience have been set by his past, and are rigidly unaffected by the actualities of the present. He is (to use the term of Gendlin and Zimring) structure-bound in his manner of experiencing. That is, he reacts "to the situation of now by finding it to be like a past experience and then reacting to that past, feeling *it*" (2). Differentiation of personal meanings in experience is crude or global, experience being seen largely in black and white terms. He does not communicate him*self*, but only communicates about externals. He tends to see himself as having no problems, or the problems he recognizes are perceived as entirely external to himself. There is much blockage of internal communication between self and experience. The individual at this stage is represented by such terms as stasis, fixity, the opposite of flow or change.

Second Stage of Process

When the person in the first stage can experience himself as fully received then the second stage follows. We seem to know very little about how to provide the experience of being received for the person in the first stage, but it is occasionally achieved in play or group therapy where the person can be exposed to a receiving climate, without himself having to take any initiative, for a long enough time to experience himself *as received.* In any event, where he does experience this, then a slight loosening and flowing of symbolic expression occurs, which tends to be characterized by the following.

Expression begins to flow in regard to non-self topics.
Example: "I guess that I suspect my father has often felt very insecure in his business relations."

Problems are perceived as external to self.
Example: "Disorganization keeps cropping up in my life."

There is no sense of personal responsibility in problems.
Example: This is illustrated in the above excerpt.

Feelings are described as unowned, or sometimes as past objects.

Example: Counselor: "If you want to tell me something of what brought you here. . . ." Client: "The symptom was — it was — just being very depressed." This is an excellent example of the way in which internal problems can be perceived and communicated about as entirely external. She is not saying "I am depressed" or even "I was depressed." Her feeling is handled as a remote, unowned object, entirely external to self.

Feelings may be exhibited, but are not recognized as such or owned.

Experiencing is bound by the structure of the past.

Example: "I suppose the compensation I always make is, rather than trying to communicate with people or have the right relationship with them, to compensate by, well, shall we say, being on an intellectual level." Here the client is beginning to recognize the way in which her experiencing is bound by the past. Her statement also illustrates the remoteness of experiencing at this level. It is as though she were holding her experience at arm's length.

Personal constructs are rigid, and unrecognized as being constructs, but are thought of as facts.

Example: "I can't ever do anything right — can't ever finish it."

Differentiation of personal meanings and feelings is very limited and global.

Example: The preceding example is a good illustration. "I can't *ever*" is one instance of a black and white differentiation, as is also the use of "right" in this absolute sense.

Contradictions may be expressed, but with little recognition of them as contradictions.

Example: "I want to know things, but I look at the same page for an hour."

As a comment on this second stage of the process of change, it might be said that a number of clients who voluntarily come for help are in this stage, but we (and probably therapists in general) have a very modest degree of success in working with them. This seems at least, to be a reasonable conclusion from Kirtner's study (5), though his conceptual framework was somewhat different. We seem to know too little about the ways in which a person at this stage may come to experience himself as "received."

Stage Three

If the slight loosening and flowing in the second stage is not blocked, but the client feels himself in these respects to be fully received as he is, then there is a still further loosening and flowing of symbolic expression. Here are some of the characteristics which seem to belong together at approximately this point on the continuum.

There is a freer flow of expression about the self as an object.

Example: "I try hard to be perfect with her — cheerful, friendly, intelligent, talkative — because I want her to love me."

There is also expression about self-related experiences as objects.

Example: "And yet there is the matter of, well, how much do you leave yourself open to marriage, and if your professional vocation is important, and that's the thing that's really yourself at this point, it does place a limitation on your contacts." In this excerpt her self is such a remote object that this would probably best be classified as being between stages two and three.

There is also expression about the self as a reflected object, existing primarily in others.

Example: "I can feel myself smiling sweetly the way my mother does, or being gruff and important the way my father does sometimes — slipping into everyone else's personalities but mine."

There is much expression about or description of feelings and personal meanings not now present.

Usually, of course, these are communications about past feelings.

Example: There were "so many things I couldn't tell people — nasty things I did. I felt so sneaky and bad."

Example: "And this feeling that came into me was just the feeling that I remember as a kid."

There is very little acceptance of feelings. For the most part feelings are revealed as something shameful, bad, or abnormal, or unacceptable in other ways.

Feelings are exhibited, and then sometimes recognized as feelings. Experiencing is described as in the past, or as somewhat remote from the self.

The preceding examples illustrate this.

Personal constructs are rigid, but may be recognized as constructs, not external facts.

Example: "I felt guilty for so much of my young life that I expect I felt I deserved to be punished most of the time anyway. If I didn't feel I deserved it for one thing, I felt I deserved it for another." Obviously he sees this as the way he has construed experience rather than as a settled fact.

Example: "I'm so much afraid wherever affection is involved it just means submission. And this I hate, but I seem to equate the two, that if I am going to get affection, then it means that I must give in to what the other person wants to do."

Differentiation of feelings and meanings is slightly sharper, less global, than in previous stages.

Example: "I mean, I was saying it before, but this time I really felt it. And is it any wonder that I felt so darn lousy when this was the way it was, that . . . they did me a dirty deal plenty of times. And conversely, I was no angel about it; I realize that."

There is a recognition of contradictions in experience.

Example: Client explains that on the one hand he has expectations of doing something great; on the other hand he feels he may easily end up as a bum.

Personal choices are often seen as ineffective.

The client "chooses" to do something, but finds that his behaviors do not fall in line with this choice.

I believe it will be evident that many people who seek psychological help are at approximately the point of stage three. They may stay at roughly this point for a considerable time describing non-present feelings and exploring the self as an object, before being ready to move to the next stage.

Stage Four

When the client feels understood, welcomed, received as he is in the various aspects of his experience at the stage three level then there is a gradual loosening of constructs, a freer flow of feelings which are characteristic of movement up the continuum. We may try to capture a number of the characteristics of this loosening, and term them the fourth phase of the process.

The client describes more intense feelings of the "not-now-present" variety.

Example: "Well, I was really — it hit me down *deep*."

Feelings are described as objects in the present.

Example: "It discourages me to feel dependent because it means I'm kind of hopeless about myself."

Occasionally feelings are expressed as in the present, sometimes breaking through almost against the client's wishes.

Example: A client, after discussing a dream including a bystander, dangerous because of having observed his "crimes," says to the therapist, "Oh, all right, I *don't* trust you."

There is a tendency toward experiencing feelings in the immediate present, and there is distrust and fear of this possibility.

Example: "I feel bound — by something or other. It must be me! There's nothing else that seems to be doing it. I can't blame it on anything else. There's this *knot* — somewhere inside of me. . . . It makes me want to get mad — and cry — and run away!"

There is little open acceptance of feelings, though some acceptance is exhibited.

The two preceding examples indicate that the client exhibits sufficient acceptance of his experience to approach some frightening feelings. But there is little conscious acceptance of them.

Experiencing is less bound by the structure of the past, is less remote, and may occasionally occur with little postponement.

Again the two preceding examples illustrate very well this less tightly bound manner of experiencing.

There is a loosening of the way experience is construed. There are some discoveries of personal constructs; there is the definite recognition of these as constructs; and there is a beginning questioning of their validity.

Example: "It amuses me. Why? Oh, because it's a little stupid of me — and I feel a little tense about it, or a little embarrassed, — and a little helpless. (*His voice softens and he looks sad.*) Humor has been my bulwark all my life; maybe it's a little out of place in trying to really look at myself. A curtain to pull down . . . I feel sort of at a loss right now. Where was I? What was I saying? I lost my grip on something — that I've been holding myself up with." Here there seems illustrated the jolting, shaking consequences of questioning a basic construct, in this case his use of humor as a defense.

There is an increased differentiation of feelings, constructs, personal meanings, with some tendency toward seeking exactness of symbolization.

Example: This quality is adequately illustrated in each of the examples in this stage.

There is a realization of concern about contradictions and incongruences between experience and self.

Example: "I'm not living up to what I am. I really should be doing more than I am. How many hours I spent on the john in this position with Mother saying, 'Don't come out 'till you've done something.' Produce! . . . That happened with lots of things."

This is both an example of concern about contradictions and a questioning of the way in which experience has been construed.

There are feelings of self responsibility in problems, though such feelings vacillate.

Though a close relationship still seems dangerous, the client risks himself, relating to some small extent on a feeling basis.

Several of the above examples illustrate this, notably the one in which the client says, "Oh, all right, I *don't* trust you."

There is no doubt that this stage and the following one constitute much of psychotherapy as we know it. These behaviors are very common in any form of therapy.

It may be well to remind ourselves again that a person is never wholly at one or another stage of the process. Listening to interviews and examining typescripts causes me to believe that a given client's expressions in a given interview may be made up, for example, of expressions and behaviors mostly characteristic of stage three, with frequent instances of rigidity characteristic of stage two or the greater loosening of stage four. It does not seem likely that one will find examples of stage six in such an interview.

The foregoing refers to the variability in the general stage of the process in which the client finds himself. If we limit ourselves to some defined area of related personal meanings in the client, then I would hypothesize much more regularity; that stage three would rarely be found before stage two; that stage four would rarely follow stage two without stage three intervening. It is this kind of tentative hypothesis which can, of course, be put to empirical test.

The Fifth Stage

As we go on up the continuum we can again try to mark a point by calling it stage five. If the client feels himself received in his expressions, behaviors, and experiences at the fourth stage then this sets in motion still further loosenings, and the freedom of organismic flow is increased. Here I believe we can again delineate crudely the qualities of this phase of the process.*

Feelings are expressed freely as in the present.

Example: "I expected kinda to get a severe rejection — this I expect all the time . . . somehow I guess I even feel it with you. . . . It's hard to talk about because I want to be the best I can possibly be with you." Here feelings regarding the therapist and the client in relationship to the therapist, emotions often most difficult to reveal, are expressed openly.

* The further we go up the scale, the less adequate are examples given in print. The reason for this is that the quality of experiencing becomes more important at these upper levels, and this can only be suggested by a transcript, certainly not fully communicated. Perhaps in time a series of recorded examples can be made available.

Feelings are very close to being fully experienced. They "bubble up," "seep through," in spite of the fear and distrust which the client feels at experiencing them with fullness and immediacy.

Example: "That kinda came out and I just don't understand it. (*Long pause*) I'm trying to get hold of what that terror is."

Example: Client is talking about an external event. Suddenly she gets a pained, stricken look.

Therapist: "What — what's hitting you now?"

Client: "I don't know. (*She cries*) . . . I must have been getting a little too close to something I didn't want to talk about, or something." Here the feeling has almost seeped through into awareness in spite of her.

Example: "I feel stopped right now. Why is my mind blank right now? I feel as if I'm hanging onto something, and I've been letting go of other things; and something in me is saying, 'What more do I have to give up?' "

There is a beginning tendency to realize that experiencing a feeling involves a direct referent.

The three examples just cited illustrate this. In each case the client knows he has experienced something, knows he is not clear as to what he has experienced. But there is also the dawning realization that the referent of these vague cognitions lies within him, in an organismic event against which he can check his symbolization and his cognitive formulations. This is often shown by expressions that indicate the closeness or distance he feels from this referent.

Example: "I really don't have my finger on it. I'm just kinda describing it."

There is surprise and fright, rarely pleasure, at the feelings which "bubble through."

Example: Client, talking about past home relationships, "That's not important any more. Hmm. (*Pause*) That was somehow very meaningful — but I don't have the slightest idea why. . . . Yes, that's it! I can forget about it now and — why, it *isn't* that important. *Wow!* All that miserableness and stuff!"

Example: Client has been expressing his hopelessness. "I'm still *amazed* at the strength of this. It seems to be *so* much the way I feel."

There is an increasing ownership of self feelings, and a desire to be these, to be the "real me."

Example: "The real truth of the matter is that I'm not the sweet, forebearing guy that I try to make out that I am. I get irritated at things. I feel like snapping at people, and I feel like being selfish at times; and I don't know why I should pretend I'm *not* that way."

This is a clear instance of the greater degree of acceptance of all feelings.

Experiencing is loosened, no longer remote, and frequently occurs with little postponement.

There is little delay between the organismic event and the full subjective living of it. A beautifully precise account of this is given by a client.

Example: "I'm still having a little trouble trying to figure out what this sadness — and the weepiness — means. I just know I feel it when I get close to a certain kind of feeling — and usually when I do get weepy, it helps me to kinda break through a wall I've set up because of things that have happened. I feel hurt about something and then automatically this kind of shields things up and then I feel like I can't really touch or feel anything very much . . . and if I'd be able to feel, or could let myself feel the instantaneous feeling when I'm hurt, I'd immediately start being weepy right then, but I can't."

Here we see him regarding his feeling as an inner referent to which he can turn for greater clarity. As he senses his weepiness he realizes that it is a delayed and partial experiencing of being hurt. He also recognizes that his defenses are such that he cannot, at this point, experience the event of hurt when it occurs.

The ways in which experience is construed are much loosened. There are many fresh discoveries of personal constructs as constructs, and a critical examination and questioning of these.

Example: A man says, "This idea of needing to please — of *having* to do it — that's really been kind of a basic assumption of my life (*he weeps quietly*). It's kind of, you know, just one of the very unquestioned axioms that I *have* to please. I have no choice. I just *have* to." Here he is clear that this assumption has been a construct, and it is evident that its unquestioned status is at an end.

There is a strong and evident tendency toward exactness in differentiation of feelings and meanings.

Example: ". . . some tension that grows in me, or some hopelessness, or some kind of incompleteness — and my life actually is very incomplete right now. . . . I just don't know. Seems to be, the closest thing it gets to, is *hopelessness*." Obviously, he is trying to capture the exact term which for him symbolizes his experience.

There is an increasingly clear facing of contradictions and incongruences in experience.

Example: "My conscious mind tells me I'm worthy. But some place inside I don't believe it. I think I'm a rat — a no-good. I've no faith in my ability to do anything."

There is an increasing quality of acceptance of self-responsibility for the problems being faced, and a concern as to how he has contributed. There are increasingly freer dialogues within the self, an improvement in and reduced blockage of internal communication.

Sometimes these dialogues are verbalized.

Example: "Something in me is saying, 'What more do I have to give up? You've taken so much from me already.' This is *me* talking to *me* — the *me* way back in there who talks to the *me* who runs the show. It's complaining now, saying, 'You're getting too close! Go away!' "

Example: Frequently these dialogues are in the form of listening to oneself, to check cognitive formulations against the direct referent of experiencing. Thus a client says, "Isn't that funny? I never really looked at it that way. I'm just trying to check it. It always seemed to me that the tension was much more externally caused than this — that it wasn't something *I used* in this way. But it's true — it's really true."

I trust that the examples I have given of this fifth phase of becoming a process will make several points clear. In the first place this phase is several hundred psychological miles from the first stage described. Here many aspects of the client are in flow, as against the rigidity of the first stage. He is very much closer to his organic being, which is always in process. He is much closer to being in the flow of his feelings. His constructions of experience are decidedly loosened and repeatedly being tested against referents and evidence within and without. Experience is much more highly differentiated, and thus internal communication, already flowing, can be much more exact.

Examples of Process in One Area

Since I have tended to speak as though the client as a whole is at one stage or another, let me stress again, before going on to describe the next stage, that in given areas of personal meaning, the process may drop below the client's general level because of experiences which are so sharply at variance with the concept of self. Perhaps I can illustrate, from a single area in the feelings of one client, something of the way the process I am describing operates in one narrow segment of experiencing.

In a case reported rather fully by Shlien (5) the quality of the self-expression in the interviews has been at approximately points three and four on our continuum of process. Then when she turns to the area of sexual problems, the process takes up at a lower level on the continuum.

In the sixth interview she feels that there are things it would be impossible to tell the therapist — then "After long pause, mentions almost inaudibly, an itching sensation in the area of the rectum, for which a physician could find no cause." Here a problem is viewed as completely external to self, the quality of experiencing is very remote. It would appear to be characteristic of the second stage of process as we have described it.

In the tenth interview, the itching has moved to her fingers. Then with great embarrassment, describes undressing games and other sex activities in childhood. Here too the quality is that of telling of nonself activities, with feelings described as past objects, though it is clearly somewhat further on the continuum of process. She concludes "because I'm just bad, dirty, that's all." Here is an expression about the self and an undifferentiated, rigid personal construct. The quality of this is that of stage three in our process, as is also the following statement about self, showing more differentiation of personal meanings. "I think inside I'm oversexed, and outside not sexy enough to attract the response I want. . . . I'd like to be the same inside and out." This last phrase has a stage four quality in its faint questioning of a personal construct.

In the twelfth interview she carries this questioning further, deciding she was not just *born* to be promiscuous. This has clearly a fourth stage quality, definitely challenging this deep-seated way of construing her experience. Also in this interview she acquires the courage to say to the therapist; "You're a man, a good looking man, and my whole problem is men like you. It would be easier if you were elderly — easier, but not better, in the long run." She is upset and embarrassed having said this and feels "it's like being naked, I'm so revealed to you." Here an immediate feeling is expressed, with reluctance and fear to be sure, but expressed, not described. Experiencing is much less remote or structure bound, and occurs with little postponement, but with much lack of acceptance. The sharper differentiation of meanings is clearly evident in the phrase "easier but not better." All of this is fully characteristic of our stage four of process.

In the fifteenth interview she describes many past experiences and feelings regarding sex, these having the quality of both the third and fourth stage as we have presented them. At some point she says, "I wanted to hurt myself, so I started going with men who would hurt me — with their penises. I enjoyed it, and was being hurt, so I had the satisfaction of being punished for my enjoyment at the same time." Here is a way of construing experience which is perceived as just that, not as an external fact. It is also quite clearly being questioned, though this questioning is implicit. There is recognition of and some concern regarding the contradictory elements in experiencing enjoyment, yet feeling she should be punished. These qualities are all fully characteristic of the fourth stage or even slightly beyond.

A bit later she describes her intense past feelings of shame at her enjoyment of sex. Her two sisters, the "neat, respected daughters" could not have orgasms, "so again I was the bad one." Up to this point this again illustrates the fourth stage. Then suddenly she asks "Or am I really lucky?" In the quality of present expression of a feeling of puzzlement, in the "bubbling through" quality, in the immediate experiencing of this wonderment, in the frank and definite questioning of her previous personal construct, this has clearly the qualities of stage five, which we have just described. She has moved forward in this process, in a climate of acceptance, a very considerable distance from stage two.

I hope this example indicates the way in which an individual, in a given area of personal meanings, becomes more and more loosened, more and more in motion, in process, as she is received. Perhaps, too, it will illustrate what I believe to be the case, that this process of increased flow is not one which happens in minutes or hours, but in weeks, or months. It is an irregularly advancing process, sometimes retreating a bit, sometimes seeming not to advance as it broadens out to cover more territory, but finally proceeding in its further flow.

The Sixth Stage

If I have been able to communicate some feeling for the scope and quality of increased loosening of feeling, experiencing and construing at each stage, then we are ready to look at the next stage which appears, from observation, to be a very crucial one. Let me see if I can convey what I perceive to be its characteristic qualities.

Assuming that the client continues to be fully received in the therapeutic relationship then the characteristics of stage five tend to be followed by a very distinctive and often dramatic phase. It is characterized as follows.

A feeling which has previously been "stuck," has been inhibited in its process quality, is experienced with immediacy now.
A feeling flows to its full result.
A present feeling is directly experienced with immediacy and richness.
This immediacy of experiencing, and the feeling which con-

stitutes its content, are accepted. This is something which is, not something to be denied, feared, struggled against.

All the preceding sentences attempt to describe slightly different facets of what is, when it occurs, a clear and definite phenomenon. It would take recorded examples to communicate its full quality, but I shall try to give an illustration without benefit of recording. A somewhat extended excerpt from the 80th interview with a young man may communicate the way in which a client comes into stage six.

Example: "I could even conceive of it as a possibility that I could have a kind of tender concern for me. . . . Still, how could *I* be tender, be concerned for *myself*, when they're one and the same thing? But yet I can *feel* it so clearly. . . . You know, like taking care of a child. You want to give it this and give it that. . . . I can kind of clearly see the purposes for somebody else . . . but I can never see them for . . . myself, that I could do this for me, you know. Is it possible that I can really want to take care of myself, and make that a major purpose of my life? That means I'd have to deal with the whole world as if I were guardian of the most cherished and most wanted possession, that this *I* was between this precious *me* that I wanted to take care of and the whole world. . . . It's almost as if I *loved* myself — you know — that's strange — but it's true."

Therapist: It seems such a strange concept to realize. Why it would mean "I would face the world as though a part of my primary responsibility was taking care of this precious individual who is me — whom I love."

Client: Whom I care for — whom I feel so *close* to. Woof! ! That's another *strange* one.

Therapist: It just seems *weird.*

Client: Yeah. It hits rather close somehow. The idea of my loving me and the taking care of me. (*His eyes grow moist.*) That's a very nice one — very nice."

The recording would help to convey the fact that here is a feeling which has never been able to flow in him, which is experienced with immediacy, in this moment. It is a feeling which flows to its full result, without inhibition. It is experienced acceptantly, with no attempt to push it to one side, or to deny it.

There is a quality of living subjectively in the experience, not feeling about it.

The client, in his *words*, may withdraw enough from the experience to feel *about* it, as in the above example, yet the recording makes it clear that his words are peripheral to the experiencing which is going on within him, and in which he is living. The best communication of this in his words is "Woof! ! That's another strange one."

Self as an object tends to disappear.

The self, at this moment, *is* this feeling. This is a being in the moment, with little self-conscious awareness, but with primarily a reflexive awareness, as Sartre terms it. The self *is*, subjectively, in the existential moment. It is not something one perceives.

Experiencing, at this stage, takes on a real process quality.

Example: One client, a man who is approaching this stage, says that he has a frightened feeling about the source of a lot of secret thoughts in himself. He goes on; "The butterflies are the thoughts closest to the surface. Underneath there's a deeper flow. I feel very removed from it all. The deeper flow is like a great school of fish moving under the surface. I see the ones that break through the surface of the water — sitting with my fishing line in one hand, with a bent pin on the end of it — trying to find a better tackle — or better yet, a way of diving in. That's the scary thing. The image I get is that *I* want to be one of the fish myself."

Therapist: "You want to be down there flowing along, too."

Though this client is not yet fully experiencing in a process manner, and hence does not fully exemplify this sixth point of the continuum, he foresees it so clearly that his description gives a real sense of its meaning.

Another characteristic of this stage of process is the physiological loosening which accompanies it.

Moistness in the eyes, tears, sighs, muscular relaxation, are frequently evident. Often there are other physiological concomitants. I would hypothesize that in these moments, had we the measure for it, we would discover improved circulation, improved conductivity of nervous impulses. An example of the "primitive" nature of some of these sensations may be indicated in the following excerpt.

Example: The client, a young man, has expressed the wish his parents would die or disappear. "It's kind of like wanting to wish them away, and wishing they had never been . . . And I'm so ashamed of myself because then they call me, and off I go — swish! They're somehow still so strong. I don't know. There's some umbilical — I can almost feel it inside me — swish (*and he gestures, plucking himself away by grasping at his navel.*)"

Therapist: "They really do have a hold on your umbilical cord."

Client: "It's funny how real it feels . . . It's like a burning sensation, kind of, and when they say something which makes me anxious I can feel it right here (*pointing*). I never thought of it quite that way."

Therapist: "As though if there's a disturbance in the relationship between you, then you do just feel it as though it was a strain on your umbilicus."

Client: "Yeah, kind of like in my gut here. It's so hard to define the feeling that I feel there."

Here he is living subjectively in the feeling of dependence on his parents. Yet it would be most inaccurate to say that he is perceiving it. He is *in* it, experiencing it as a strain on his umbilical cord. *In this stage, internal communication is free and relatively unblocked.*

I believe this is quite adequately illustrated in the examples given. Indeed the phrase, "internal communication" is no longer quite correct, for as each of these examples illustrates, the crucial moment is a moment of integration, in which communication between different internal foci is no longer necessary, because they become *one.*

The incongruence between experience and awareness is vividly experienced as it disappears into congruence.
The relevant personal construct is dissolved in this experiencing moment, and the client feels cut loose from his previously stabilized framework.

I trust these two characteristics may acquire more meaning from the following example. A young man has been having difficulty getting close to a certain unknown feeling. "That's almost exactly what the feeling is, too — it was that I was living so much of my life, and seeing so much of my life in terms of being scared of something." He tells how his professional activities are just to give him a little safety and "a little world where I'll be secure, you know. And for the same reason. (*Pause*) I was kind of letting it seep through. But I also tied it in with you and with my relationship with you, and one thing I feel about it is fear of its going away. (*His tone changes to role-play more accurately his feeling.*) Won't you *let* me have this? I kind of *need* it. I can be so lonely and scared without it."

Therapist: "M-hm, m-hm. 'Let me hang on to it because I'd be terribly scared if I didn't! . . .' It's a kind of pleading thing too, isn't it?"

Client: "I get a sense of — it's this kind of pleading little boy. It's this gesture of begging. (*Putting his hands up as if in prayer.*)

Therapist: "You put your hands in kind of a supplication."

Client: "Yeah, that's right. '*Won't* you do this for me?' kind of. Oh, that's terrible! Who, Me? Beg? . . . That's an emotion I've never felt clearly at all — something I've never been . . . (*Pause*) . . . I've got such a confusing feeling. One is, it's such a wondrous feeling to have these new things come out of me. It amazes me so much each time, and there's that same feeling, being scared that I've so much of this. (*Tears*) . . . I just don't know myself. Here's suddenly something I never realized, hadn't any inkling of — that it was some *thing* or *way* I wanted to be."

Here we see a complete experiencing of his pleadingness, and a vivid recognition of the discrepancy between this experiencing and his concept of himself. Yet this experiencing of discrepancy exists in the moment of its disappearance. From now on he *is* a person who feels *pleading*, as well as many other feelings. As this moment dissolves the way he has construed himself he feels cut loose from his previous world — a sensation which is both wondrous and frightening.

The moment of full experiencing becomes a clear and definite referent.

The examples given should indicate that the client is often not too clearly aware of what has "hit him" in these moments. Yet this does not seem too important because the event is an entity, a referent, which can be returned to, again and again, if necessary, to discover more about it. The pleadingness, the feeling of "loving myself" which are present in these examples, may not prove to be exactly as described. They are, however, solid points of reference to which the client can return until he has satisfied himself as to what they are. It is, perhaps, that they constitute a clear-cut physiological event, a substratum of the conscious life, which the client can return to for investigatory purposes. Gendlin has called my attention to this significant quality of experiencing as a referent. He is endeavoring to build an extension of psychological theory on this basis. (1)

Differentiation of experiencing is sharp and basic.

Because each of these moments is a referent, a specific entity, it does not become confused with anything else. The process of sharp differentiation builds on it and about it.

In this stage, there are no longer "problems," external or internal. The client is living, subjectively, a phase of his problem. It is not an object.

I trust it is evident that in any of these examples, it would be grossly inaccurate to say that the client perceives his problem as internal, or is dealing with it as an internal problem. We need some way of indicating that he is further than this, and of course enormously far in the process sense from perceiving his problem as external. The best description seems to be that he neither perceives his problem nor deals with it. He is simply living some portion of it knowingly and acceptingly.

I have dwelt so long on this sixth definable point on the process continuum because I see it as a highly crucial one. My observation is that these moments of immediate, full, accepted experiencing are in some sense almost irreversible. To put this in terms of the examples, it is my observation and hypothesis that with these clients, whenever a future experiencing of the same quality and characteristics occurs, it will necessarily be recognized in awareness for what it is: a tender caring for self, an umbilical bond which makes him a part of his parents, or a pleading small-boy dependence, as the case may be. And, it might be remarked in passing, once an experience is fully in awareness, fully accepted, then it can be coped with effectively, like any other clear reality.

The Seventh Stage

In those areas in which the sixth stage has been reached, it is no longer so necessary that the client be fully received by the therapist, though this still seems helpful. However, because of the tendency for the sixth stage to be irreversible, the client often seems to go on into the seventh and final stage without much need of the therapist's help. This stage occurs as much outside of the therapeutic relationship as in it, and is often reported, rather than experienced in the therapeutic hour. I shall try to describe some of its characterictics as I feel I have observed them.

New feelings are experienced with immediacy and richness of detail, both in the therapeutic relationship and outside.
The experiencing of such feelings is used as a clear referent.

The client quite consciously endeavors to use these referents in order to know in a clearer and more differentiated way who he is, what he wants, and what his attitudes are. This is true even when the feelings are unpleasant or frightening.

There is a growing and continuing sense of acceptant ownership of these changing feelings, a basic trust in his own process.

This trust is not primarily in the conscious processes which go on, but rather in the total organismic process. One client describes the way in which experience characteristic of the sixth stage looks to him, describing it in terms characteristic of the seventh stage.

"In therapy here, what has counted is sitting down and saying, 'this is what's bothering me,' and play around with it for awhile until something gets squeezed out through some emotional crescendo, and the thing is over with — looks different. Even then, I can't tell just exactly what's happened. It's just that I exposed something, shook it up and turned it around; and when I put it back it felt better. It's a little frustrating because I'd like to know exactly what's going on. . . . This is a funny thing because it feels as if I'm not doing anything at all about it — the only *active* part I take is to — to be alert and grab a thought as it's going by . . . And there's sort of a feeling, 'Well now, what will I do with it, now that I've seen it right?' There's no handles on it you can adjust or anything. Just talk about it awhile, and let it go. And apparently that's all there is to it. Leaves me with a somewhat unsatisfied feeling though — a feeling that I haven't accomplished anything. It's been accomplished without my knowledge or consent. . . . The point is I'm not sure of the quality of the readjustment because I didn't get to see it, to check on it. . . . All I can do is observe the facts — that I look at things a little differently and am less anxious, by a long shot, and a lot more active. Things are looking up in general. I'm very happy with the way things have gone. But I feel sort of like a spectator." A few moments later, following this rather grudging acceptance of the process going on in him, he adds, "I seem to work best when my conscious mind is only concerned with facts and letting the analysis of them go on by itself without paying any attention to it."

Experiencing has lost almost completely its structure-bound aspects and becomes process experiencing — that is, the situation is experienced and interpreted in its newness, not as the past.

The example given in stage six suggests the quality I am trying to describe. Another example in a very specific area is given by a client in a follow-up interview as he explains the different quality that has come about in his creative work. It used to be that he tried to be orderly. "You begin at the beginning and you progress regularly through to the end." Now he is aware that the process in himself is different. "When I'm working on an idea, the whole idea develops like the latent image coming out when you develop a photograph. It doesn't start at one edge and fill in over to the other. It comes in *all over*. At first all you see is the hazy outline, and you wonder what it's going to be; and then gradually something fits here and something fits there, and pretty soon it all comes clear — all at once." It is obvious that he has not only come to trust this process, but that he is experiencing it as it *is*, not in terms of some past.

The self becomes increasingly simply the subjective and reflexive awareness of experiencing. The self is much less frequently a perceived object, and much more frequently something confidently felt in process.

An example may be taken from the same follow-up interview with the client quoted above. In this interview, because he is reporting his experience since therapy, he again becomes aware of himself as an object, but it is clear that this has not been the quality of his day-by-day experience. After reporting many changes, he says, "I hadn't really thought of any of these things in connection with therapy until tonight. . . . (*Jokingly*) Gee! maybe something *did* happen. Because my life since has been different. My productivity has gone up. My confidence has gone up. I've become brash in situations I would have avoided before. And also, I've become much less brash in situations where I would have become very obnoxious before." It is clear that only afterward does he realize what his self has been.

Personal constructs are tentatively reformulated, to be validated against further experience, but even then, to be held loosely.

A client describes the way in which such a construct changed, between interviews, toward the end of therapy.

"I don't know what (changed), but I definitely feel different about looking back at my childhood, and some of the hostility about my mother and father has evaporated. I substituted for a feeling of resentment about them a sort of acceptance of the fact that they did a number of things that were undesirable with me. But I substituted a sort of feeling of interested excitement that — gee — now that I'm finding out what was wrong, *I* can do something about it — correct their mistakes." Here the way in which he construes his experience with his parents has been sharply altered.

Another example may be taken from an interview with a client who has always felt that he had to please people. "I can see . . . what it would be like — that it doesn't matter if I don't please you — that pleasing you or not pleasing you is not the thing that is important to me. If I could just kinda say that to people — you know? . . . the idea of just spontaneously saying something — and it not mattering whether it pleases or not — Oh God! you could say almost *anything:* But that's true, you know." And a little later he asks himself, with incredulity, "You mean if I'd really be what I *feel* like being, that that would be all right?" He is struggling toward a reconstruing of some very basic aspects of his experience.

Internal communication is clear, with feelings and symbols well matched, and fresh terms for new feelings.
There is the experiencing of effective choice of new ways of being.

Because all the elements of experience are available to awareness, choice becomes real and effective. Here a client is just coming to this realization. "I'm trying to encompass a way of talking that is a way out of being scared of talking. Perhaps just kind of thinking out loud is the way to do that. But I've got so *many* thoughts I could only do it a little bit. But maybe I could let my talk be an expression of my real thoughts, instead of just trying to make the proper noises in each situation." Here he is sensing the possibility of effective choice.

Another client comes in telling of an argument he had with his wife. "I wasn't so angry with myself. I didn't hate myself so much. I realized 'I'm acting childishly' and somehow I chose to do that."

It is not easy to find examples by which to illustrate this seventh stage, because relatively few clients fully achieve this point. Let me try to summarize briefly the qualities of this end point of the continuum.

When the individual has, in his process of change, reached the seventh stage, we find ourselves involved in a new dimension. The client has now incorporated the quality of motion, of flow, of changingness, into every aspect of his psychological life, and this becomes its outstanding characteristic. He lives in his feelings, knowingly and with basic trust in them and acceptance of them. The ways in which he construes experience are continually changing as his personal constructs are modified by each new living event. His experiencing is process in nature, feeling the new in each situation and interpreting it anew, interpreting in terms of the past only to the extent that the now is identical with the past. He experiences with a quality of immediacy, knowing at the same time *that* he experiences. He values exactness in differentiation of his feelings and of the personal meanings of his experience. His internal communication between various aspects of himself is free and unblocked. He communicates himself freely in relationships with others, and these relationships are not stereotyped, but person to person. He is aware of himself, but not as an object. Rather it is a reflexive awareness, a subjective living in himself in motion. He perceives himself as responsibly related to his problems. Indeed, he feels a fully responsible relationship to his life in all its fluid aspects. He lives fully in himself as a constantly changing flow of process.

Some Questions Regarding This Process Continuum

Let me try to anticipate certain questions which may be raised about the process I have tried to describe.

Is this *the* process by which personality changes or one of many kinds of change? This I do not know. Perhaps there are several types of process by which personality changes. I would only specify that this seems to be the process which is set in motion when the individual experiences himself as being fully received.

Does it apply in all psychotherapies, or is this the process which occurs in one psychotherapeutic orientation only? Until we have more recordings of therapy from other orientations, this question cannot be answered. However, I would hazard a guess that perhaps therapeutic approaches which place great stress on the cognitive and little on the emotional aspects of experience may set in motion an entirely different process of change.

Would everyone agree that this is a desirable process of change, that it moves in valued directions? I believe not. I believe some people do not value fluidity. This will be one of the social value judgments which individuals and cultures will have to make. Such a process of change can easily be avoided, by reducing or avoiding those relationships in which the individual is fully received as he is.

Is change on this continuum rapid? My observation is quite the contrary. My interpretation of Kirtner's study (4), which may be slightly different from his, is that a client might start therapy at about stage two and end at about stage four with both client and therapist being quite legitimately satisfied that substantial progress had been made. It would occur very rarely, if ever, that a client who fully exemplified stage one would move to a point where he fully exemplified stage seven. If this did occur, it would involve a matter of years.

Are the descriptive items properly grouped at each stage? I feel sure that there are many errors in the way I have grouped my observations. I also wonder what important elements have been omitted. I wonder also if the different elements of this continuum might not be more parsimoniously described. All such questions, however, may be given an empirical answer, if the hypothesis I am setting forth has merit in the eyes of various research workers.

SUMMARY

I have tried to sketch, in a crude and preliminary manner, the flow of a process of change which occurs when a client experiences himself as being received, welcomed, understood as he is. This process involves several threads, separable at first, becoming more of a unity as the process continues.

This process involves a loosening of feelings. At the lower end of the continuum they are described as remote, unowned, and not now present. They are then described as present objects with some sense of ownership by the individual. Next they are expressed as owned feelings in terms closer to their immediate experiencing. Still further up the scale they are experienced and expressed in the immediate present with a decreasing fear of this process. Also, at this point, even those feelings which have been previously denied to awareness bubble through into awareness, are experienced, and increasingly owned. At the upper end of the continuum living in the process of experiencing a continually changing flow of feelings becomes characteristic of the individual.

The process involves a change in the manner of experiencing. The continuum begins with a fixity in which the individual is very remote from his experiencing and unable to draw upon or symbolize its implicit meaning. Experiencing must be safely in the past before a meaning can be drawn from it and the present is interpreted in terms of these past meanings. From this remoteness in relation to his experiencing, the individual moves toward the recognition of experiencing as a troubling process going on within him. Experiencing gradually becomes a more accepted inner referent to which he can turn for increasingly accurate meanings. Finally he becomes able to live freely and acceptantly in a fluid process of experiencing, using it comfortably as a major reference for his behavior.

The process involves a shift from incongruence to congruence. The continuum runs from a maximum of incongruence which is quite unknown to the individual through stages where there is an increasingly sharp recognition of the contradictions and discrepancies existing within himself to the experiencing of incongruence in the immediate present in a way which dissolves this. At the upper end of the continuum, there would never be more than temporary incongruence between experiencing and awareness since the individual would not need to defend himself against the threatening aspects of his experience.

The process involves a change in the manner in which, and the extent to which the individual is able and willing to communicate himself in a receptive climate. The continuum runs from a complete unwillingness to communicate self to the self as a rich and changing awareness of internal experiencing which is readily communicated when the individual desires to do so.

The process involves a loosening of the cognitive maps of experience. From construing experience in rigid ways which are perceived as external facts, the client moves toward developing changing, loosely held construings of meaning in experience, constructions which are modifiable by each new experience.

There is a change in the individual's relationship to his problems. At one end of the continuum problems are unrecognized and there is no desire to change. Gradually there is a recognition that problems exist. At a further stage, there is recognition that the individual has contributed to these problems, that they have not arisen entirely from external sources. Increasingly, there is a sense of self-responsibility for the problems. Further up the continuum there is a living or experiencing of some aspect of the problems. The person lives his problems subjectively, feeling responsible for the contribution he has made in the development of his problems.

There is change in the individual's manner of relating. At one end of the continuum the individual avoids close relationships, which are perceived as being dangerous. At the other end of the continuum, he lives openly and freely in relation to the therapist and to others, guiding his behavior in the relationship on the basis of his immediate experiencing.

In general, the process moves from a point of fixity, where all the elements and threads described above are separately discernible and separately understandable, to the flowing peak moments of therapy in which all these threads become inseparably woven together. In the new experiencing with immediacy which occurs at such moments, feeling and cognition interpenetrate, self is subjectively present in the experience, volition is simply the subjective following of a harmonious balance of organismic direction. Thus, as the process reaches this point the person becomes a unity of flow, of motion. He has changed, but what seems most significant, he has become an integrated process of changingness.

REFERENCES

1. Gendlin, E. *Experiencing and the Creation of Meaning* (tentative title). Glencoe, Ill.: Free Press. (In Press) (Especially Chap. 7)

2. Gendlin, E., and F. Zimring. The qualities or dimensions of experiencing and their change. *Counseling Center Discussion Papers 1*, #3, Oct. 1955. University of Chicago Counseling Center.

3. Kelly, G. A. *The psychology of personal constructs*. Vol. 1. New York: Norton, 1955.

4. Kirtner, W. L., and D. S. Cartwright. Success and failure in client-centered therapy as a function of initial in-therapy behavior. *J. Consult. Psychol.*, 1958, *22*, 329–333.

5. Lewis, M. K., C. R. Rogers, and John M. Shlien. Two cases of time-

limited client-centered psychotherapy. In Burton, A. (Ed.), *Case Studies of Counseling and Psychotherapy*. New York: Prentice-Hall, 1959, 309–352.

6. Mooney, R. L. The researcher himself. In *Research for curriculum improvement*. Nat'l Educ. Ass'n., 1957, Chap. 7.

7. Mooney, R. L. Problems in the development of research men. *Educ. Research Bull.*, *30*, 1951, 141–150.

8. Rogers, C. R. The necessary and sufficient conditions of therapeutic personality change. *J. Consult. Psychol.*, 1957, *21*, 95–103.

PART IV

A Philosophy of Persons

I have formed some philosophical impressions
of the life and goal toward which
the individual moves when he is free.

8

"To Be That Self Which One Truly Is"
A Therapist's View of Personal Goals

In these days most psychologists regard it as an insult if they are accused of thinking philosophical thoughts. I do not share this reaction. I cannot help but puzzle over the meaning of what I observe. Some of these meanings seem to have exciting implications for our modern world.

In 1957 Dr. Russell Becker, a friend, former student and colleague of mine, invited me to give a special lecture to an all-college convocation at Wooster College in Ohio. I decided to work out more clearly for myself the meaning of the personal directions which clients seem to take in the free climate of the therapeutic relationship. When the paper was finished I had grave doubts that I had expressed anything which was in any way new or significant. The rather astonishingly long-continued applause of the audience relieved my fears to some degree.

As the passage of time has enabled me to look more objectively at what I said, I feel satisfaction on two counts. I believe it expresses well the observations which for me have crystallized into two important themes: my confidence in the human organism, when it is functioning freely; and the existential quality of satisfying living, a theme presented by some of our most modern philosophers, which was however beautifully expressed more than twenty-five centuries ago by Lao-tzu, when he said, "The way to do is to be."

THE QUESTIONS

"What is my goal in life?" "What am I striving for?" "What is my purpose?" These are questions which every individual asks himself at one time or another, sometimes calmly and meditatively, sometimes in agonizing uncertainty or despair. They are old, old questions which have been asked and answered in every century of history. Yet they are also questions which every individual must ask and answer for himself, in his own way. They are questions which I, as a counselor, hear expressed in many differing ways as men and women in personal distress try to learn, or understand, or choose, the directions which their lives are taking.

In one sense there is nothing new which can be said about these questions. Indeed the opening phrase in the title I have chosen for this paper is taken from the writings of a man who wrestled with these questions more than a century ago. Simply to express another personal opinion about this whole issue of goals and purposes would seem presumptuous. But as I have worked for many years with troubled and maladjusted individuals I believe that I can discern a pattern, a trend, a commonality, an orderliness, in the tentative answers to these questions which they have found for themselves. And so I would like to share with you my perception of what human beings appear to be striving for, when they are free to choose.

SOME ANSWERS

Before trying to take you into this world of my own experience with my clients, I would like to remind you that the questions I have mentioned are not pseudo-questions, nor have men in the past or at the present time agreed on the answers. When men in the past have asked themselves the purpose of life, some have answered, in the words of the catechism, that "the chief end of man is to glorify God." Others have thought of life's purpose as being the preparation of oneself for immortality. Others have settled on a much more earthy goal — to enjoy and release and satisfy every sensual desire. Still others — and this applies to many today — regard the purpose of life as being to achieve — to gain material possessions, status, knowledge, power. Some have made it their goal to give themselves completely and devotedly to a cause outside of themselves such as Christianity, or Communism. A Hitler has seen his goal as that of becoming the leader of a master race which would exercise power over all. In sharp contrast, many an Oriental has striven to eliminate all personal desires, to exercise the utmost of control over himself. I mention these widely ranging choices to indicate some of the very different aims men have lived for, to suggest that there are indeed many goals possible.

In a recent important study Charles Morris investigated objectively the pathways of life which were preferred by students in six different countries — India, China, Japan, the United States, Canada, and Norway (5). As one might expect, he found decided differences in goals between these national groups. He also endeavored, through a factor analysis of his data, to determine the underlying dimensions of value which seemed to operate in the thousands of specific individual preferences. Without going into the details of his analysis, we might look at the five dimensions which emerged, and which, combined in various positive and negative ways, appeared to be responsible for the individual choices.

The first such value dimension involves a preference for a responsible, moral, self-restrained participation in life, appreciating and conserving what man has attained.

The second places stress upon delight in vigorous action for the overcoming of obstacles. It involves a confident initiation of change, either in resolving personal and social problems, or in overcoming obstacles in the natural world.

The third dimension stresses the value of a self-sufficient inner life with a rich and heightened self-awareness. Control over persons and things is rejected in favor of a deep and sympathetic insight into self and others.

The fourth underlying dimension values a receptivity to persons and to nature. Inspiration is seen as coming from a source outside the self, and the person lives and develops in devoted responsiveness to this source.

The fifth and final dimension stresses sensuous enjoyment, self-enjoyment. The simple pleasures of life, an abandonment to the moment, a relaxed openness to life, are valued.

This is a significant study, one of the first to measure objectively the answers given in different cultures to the question, what is the purpose of my life? It has added to our knowledge of the answers given. It has also helped to define some of the basic dimensions in terms of which the choice is made. As Morris says, speaking of these dimensions, "it is as if persons in various cultures have in common five major tones in the musical scales on which they compose different melodies." (5, p. 185)

ANOTHER VIEW

I find myself, however, vaguely dissatisfied with this study. None of the "Ways to Live" which Morris put before the students as possible choices, and none of the factor dimensions, seems to contain satisfactorily the goal of life which emerges in my experience with my clients. As I watch person after person struggle in his therapy hours to find a way of life for himself, there seems to be a general pattern emerging, which is not quite captured by any of Morris' descriptions.

The best way I can state this aim of life, as I see it coming to light in my relationship with my clients, is to use the words of Søren Kierkegaard — "to be that self which one truly is." (3, p. 29) I am quite aware that this may sound so simple as to be absurd. To be what one is seems like a statement of obvious fact rather than a goal. What does it mean? What does it imply? I want to devote the remainder of my remarks to those issues. I will simply say at the outset that it seems to mean and imply some strange things. Out of my experience with my clients, and out of my own self-searching, I find myself arriving at views which would have been very foreign to me ten or fifteen years ago. So I trust you will look at these views with critical scepticism, and accept them only in so far as they ring true in your own experience.

Directions Taken by Clients

Let me see if I can draw out and clarify some of the trends and tendencies which I see as I work with clients. In my relationship with these individuals my aim has been to provide a climate which contains as much of safety, of warmth, of empathic understanding, as I can genuinely find in myself to give. I have not found it satisfying or helpful to intervene in the client's experience with diagnostic or interpretative explanations, nor with suggestions and guidance. Hence the trends which I see appear to me to come from the client himself, rather than emanating from me.*

Away From Façades

I observe first that characteristically the client shows a tendency to move away, hesitantly and fearfully, from a self that he is *not*. In other words even though there may be no recognition of what he might be moving toward, he is moving away from something. And of course in so doing he is beginning to define, however negatively, what he *is*.

At first this may be expressed simply as a fear of exposing what he is. Thus one eighteen-year-old boy says, in an early interview: "I know I'm not so hot, and I'm afraid they'll find it out. That's why I do these things. . . . They're going to find out some day that I'm not so hot. I'm just trying to put that day off as long as possible. . . . If you know me as I know myself —. (*Pause*) I'm not going to tell you the person I really think I am. There's only one place I won't cooperate and that's it. . . . It wouldn't help your opinion of me to know what I think of myself."

It will be clear that the very expression of this fear is a part of becoming what he is. Instead of simply *being* a façade, as if it were himself, he is coming closer to being *himself*, namely a frightened

* I cannot close my mind, however, to the possibility that someone might be able to demonstrate that the trends I am about to describe might in some subtle fashion, or to some degree, have been initiated by me. I am describing them as occurring in the client in this safe relationship, because that seems the most likely explanation.

person hiding behind a façade because he regards himself as too awful to be seen.

AWAY FROM "OUGHTS"

Another tendency of this sort seems evident in the client's moving away from the compelling image of what he "ought to be." Some individuals have absorbed so deeply from their parents the concept "I ought to be good," or "I have to be good," that it is only with the greatest of inward struggle that they find themselves moving away from this goal. Thus one young woman, describing her unsatisfactory relationship with her father, tells first how much she wanted his love. "I think in all this feeling I've had about my father, that *really* I *did* very much want a good relationship with him. . . . I wanted so much to have him care for me, and yet didn't seem to get what I really wanted." She always felt she had to meet all of his demands and expectations and it was "just too much. Because once I meet one there's another and another and another, and I never really meet them. It's sort of an endless demand." She feels she has been like her mother, submissive and compliant, trying continually to meet his demands. "And really *not* wanting to be that kind of person. I find it's not a good way to be, but yet I think I've had a sort of belief that that's the way you *have* to be if you intend to be thought a lot of and loved. And yet who would *want* to love somebody who was that sort of wishy washy person?" The counselor responded, "Who really would love a door mat?" She went on, "At least I wouldn't want to be loved by the kind of person who'd love a door mat!"

Thus, though these words convey nothing of the self she might be moving toward, the weariness and disdain in both her voice and her statement make it clear that she is moving away from a self which *has* to be good, which *has* to be submissive.

Curiously enough a number of individuals find that they have felt compelled to regard themselves as bad, and it is this concept of themselves that they find they are moving away from. One young man shows very clearly such a movement. He says: "I don't know how I got this impression that being ashamed of myself was such an *appropriate* way to feel. . . . Being ashamed of me was the way

I just *had* to be. . . . There was a world where being ashamed of myself was the best way to feel. . . . If you are something which is disapproved of very much, then I guess the only way you can have any kind of self-respect is to be ashamed of that part of you which isn't approved of. . . .

"But now I'm adamantly refusing to do things from the old viewpoint. . . . It's as if I'm convinced that someone said, 'The way you will *have* to be is to be *ashamed* of yourself — so *be* that way!' And I accepted it for a long, long time, saying 'OK, that's me!' And now I'm standing up against that somebody, saying, 'I don't care *what* you say. I'm *not* going to feel ashamed of myself!' " Obviously he is abandoning the concept of himself as shameful and bad.

Away From Meeting Expectations

Other clients find themselves moving away from what the culture expects them to be. In our current industrial culture, for example, as Whyte has forcefully pointed out in his recent book (7), there are enormous pressures to become the characteristics which are expected of the "organization man." Thus one should be fully a member of the group, should subordinate his individuality to fit into the group needs, should become "the well-rounded man who can handle well-rounded men."

In a newly completed study of student values in this country Jacob summarizes his findings by saying, "The main overall effect of higher education upon student values is to bring about general acceptance of a body of standards and attitudes characteristic of collegebred men and women in the American community. . . . The impact of the college experience is . . . to *socialize* the individual, to refine, polish, or 'shape up' his values so that he can fit comfortably into the ranks of American college alumni." (1, p. 6)

Over against these pressures for conformity, I find that when clients are free to be any way they wish, they tend to resent and to question the tendency of the organization, the college or the culture to mould them to any given form. One of my clients says with considerable heat: "I've been so long trying to live according to what was meaningful to other people, and what made no sense at *all* to me, really. I somehow felt so much *more* than that, at some level."

So he, like others, tends to move away from being what is expected.

AWAY FROM PLEASING OTHERS

I find that many individuals have formed themselves by trying to please others, but again, when they are free, they move away from being this person. So one professional man, looking back at some of the process he has been through, writes, toward the end of therapy: "I finally felt that I simply *had* to begin doing what *I wanted* to do, not what I thought I *should* do, and regardless of what other people feel I *should* do. This is a complete reversal of my whole life. I've always felt I *had* to do things because they were expected of me, or more important, to make people like me. The hell with it! I think from now on I'm going to just be me — rich or poor, good or bad, rational or irrational, logical or illogical, famous or infamous. So thanks for your part in helping me to rediscover Shakespeare's — 'To thine own *self* be true.' "

So one may say that in a somewhat negative way, clients define their goal, their purpose, by discovering, in the freedom and safety of an understanding relationship, some of the directions they do *not* wish to move. They prefer not to hide themselves and their feelings from themselves, or even from some significant others. They do not wish to be what they "ought" to be, whether that imperative is set by parents, or by the culture, whether it is defined positively or negatively. They do not wish to mould themselves and their behavior into a form which would be merely pleasing to others. They do not, in other words, choose to be anything which is artificial, anything which is imposed, anything which is defined from without. They realize that they do not value such purposes or goals, even though they may have lived by them all their lives up to this point.

TOWARD SELF-DIRECTION

But what is involved positively in the experience of these clients? I shall try to describe a number of the facets I see in the directions in which they move.

First of all, the client moves toward being autonomous. By this I mean that gradually he chooses the goals toward which *he* wants to move. He becomes responsible for himself. He decides what activities and ways of behaving have meaning for him, and what do not. I think this tendency toward self-direction is amply illustrated in the examples I have given.

I would not want to give the impression that my clients move blithely or confidently in this direction. No indeed. Freedom to be oneself is a frighteningly responsible freedom, and an individual moves toward it cautiously, fearfully, and with almost no confidence at first.

Nor would I want to give the impression that he always makes sound choices. To be responsibly self-directing means that one chooses — and then learns from the consequences. So clients find this a sobering but exciting kind of experience. As one client says — "I feel frightened, and vulnerable, and cut loose from support, but I also feel a sort of surging up or force or strength in me." This is a common kind of reaction as the client takes over the self-direction of his own life and behavior.

Toward Being Process

The second observation is difficult to make, because we do not have good words for it. Clients seem to move toward more openly being a process, a fluidity, a changing. They are not disturbed to find that they are not the same from day to day, that they do not always hold the same feelings toward a given experience or person, that they are not always consistent. They are in flux, and seem more content to continue in this flowing current. The striving for conclusions and end states seems to diminish.

One client says, "Things are sure changing, boy, when I can't even predict my own behavior in here anymore. It was something I was able to do before. Now I don't know what I'll say next. Man, it's quite a feeling. . . . I'm just surprised I even said these things. . . . I see something new every time. It's an adventure, that's what it is — into the unknown. . . . I'm beginning to enjoy this now, I'm joyful about it, even about all these old negative things." He is beginning to appreciate himself as a fluid process, at first in the therapy hour, but later he will find this true in his life. I cannot help but be reminded of Kierkegaard's description of the individual who really exists. "An existing individual is constantly in process of becoming, . . . and translates all his thinking into terms of process. It is with (him) . . . as it is with a writer and his style; for he only has a style who never has anything finished, but 'moves the waters of the language' every time he begins, so that the most common expression comes into being for him with the freshness of a new birth." (2, p. 79) I find this catches excellently the direction in which clients move, toward being a process of potentialities being born, rather than being or becoming some fixed goal.

Toward Being Complexity

It also involves being a complexity of process. Perhaps an illustration will help here. One of our counselors, who has himself been much helped by psychotherapy, recently came to me to discuss his relationship with a very difficult and disturbed client. It interested me that he did not wish to discuss the client, except in the briefest terms. Mostly he wanted to be sure that he was clearly aware of the complexity of his own feelings in the relationship — his warm feelings toward the client, his occasional frustration and annoyance, his sympathetic regard for the client's welfare, a degree of fear that the client might become psychotic, his concern as to what others would think if the case did not turn out well. I realized that his overall attitude was that if he could *be*, quite openly and transparently, all of his complex and changing and sometimes contradictory feelings in the relationship, all would go well. If, however, he was only part of his feelings, and partly façade or defense, he was sure the relationship would not be good. I find that this desire to be *all* of oneself in each moment — all the richness and complexity, with nothing hidden from oneself, and nothing feared in oneself — this is a common desire in those who have seemed to show much movement in therapy. I do not need to say that this is a difficult, and in its absolute sense an impossible goal. Yet one of the most evident trends in clients is to move toward becoming all of the complexity of one's changing self in each significant moment.

Toward Openness to Experience

"To be that self which one truly is" involves still other components. One which has perhaps been implied already is that the individual moves toward living in an open, friendly, close relationship to his own experience. This does not occur easily. Often as the client senses some new facet of himself, he initially rejects it. Only as he experiences such a hitherto denied aspect of himself in an acceptant climate can he tentatively accept it as a part of himself. As one client says with some shock after experiencing the dependent, small boy aspect of himself, "That's an emotion I've never felt clearly — one that I've never been!" He cannot tolerate the experience of his childish feelings. But gradually he comes to accept and embrace them as a part of himself, to live close to them and in them when they occur.

Another young man, with a very serious stuttering problem, lets himself be open to some of his buried feelings toward the end of his therapy. He says, "Boy, it was a terrible fight. I never realized it. I guess it was too painful to reach that height. I mean I'm just beginning to feel it now. Oh, the *terrible* pain. . . . It was *terrible* to talk. I mean I wanted to talk and then I didn't want to. . . . I'm feeling — I think I know — it's just plain strain — terrible strain — *stress*, that's the word, just so much *stress* I've been feeling. I'm just beginning to *feel* it now after all these years of it. . . . it's terrible. I can hardly get my breath now too, I'm just all choked up inside, all *tight* inside. . . . I just feel like I'm *crushed*. (*He begins to cry.*) I never realized that, I never knew that." (6) Here he is opening himself to internal feelings which are clearly not new to him, but which up to this time, he has never been able fully to experience. Now that he can permit himself to experience them, he will find them less terrible, and he will be able to live closer to his own experiencing.

Gradually clients learn that experiencing is a friendly resource, not a frightening enemy. Thus I think of one client who, toward the close of therapy, when puzzled about an issue, would put his head in his hands and say, "Now what *is* it I'm feeling? I want to get next to it. I want to learn what it is." Then he would wait, quietly and patiently, until he could discern the exact flavor of the feelings occurring in him. Often I sense that the client is trying to listen to himself, is trying to hear the messages and meanings which are being communicated by his own physiological reactions. No longer is he so fearful of what he may find. He comes to realize that his own inner reactions and experiences, the messages of his senses and his viscera, are friendly. He comes to want to be close to his inner sources of information rather than closing them off.

Maslow, in his study of what he calls self-actualizing people, has noted this same characteristic. Speaking of these people, he says, "Their ease of penetration to reality, their closer approach to an animal-like or child-like acceptance and spontaneity imply a superior awareness of their own impulses, their own desires, opinions, and subjective reactions in general." (4, p. 210)

This greater openness to what goes on within is associated with a similar openness to experiences of external reality. Maslow might be speaking of clients I have known when he says, "self-actualized people have a wonderful capacity to appreciate again and again, freshly and naively, the basic goods of life with awe, pleasure, wonder, and even ecstasy, however stale these experiences may be for other people." (4, p. 214)

TOWARD ACCEPTANCE OF OTHERS

Closely related to this openness to inner and outer experience in general is an openness to and an acceptance of other individuals. As a client moves toward being able to accept his own experience, he also moves toward the acceptance of the experience of others. He values and appreciates both his own experience and that of others for what it *is*. To quote Maslow again regarding his self-actualizing individuals: "One does not complain about water because it is wet, nor about rocks because they are hard. . . . As the child looks out upon the world with wide, uncritical and innocent eyes, simply noting and observing what is the case, without either arguing the matter or demanding that it be otherwise, so does the self-actualizing person look upon human nature both in himself and in others." (4, p. 207) This acceptant attitude toward that which exists, I find developing in clients in therapy.

Toward Trust of Self

Still another way of describing this pattern which I see in each client is to say that increasingly he trusts and values the process which is himself. Watching my clients, I have come to a much better understanding of creative people. El Greco, for example, must have realized as he looked at some of his early work, that "good artists do not paint like that." But somehow he trusted his own experiencing of life, the process of himself, sufficiently that he could go on expressing his own unique perceptions. It was as though he could say, "Good artists do not paint like this, but *I* paint like this." Or to move to another field, Ernest Hemingway was surely aware that "good writers do not write like this." But fortunately he moved toward being Hemingway, being himself, rather than toward some one else's conception of a good writer. Einstein seems to have been unusually oblivious to the fact that good physicists did not think his kind of thoughts. Rather than drawing back because of his inadequate academic preparation in physics, he simply moved toward being Einstein, toward thinking his own thoughts, toward being as truly and deeply himself as he could. This is not a phenomenon which occurs only in the artist or the genius. Time and again in my clients, I have seen simple people become significant and creative in their own spheres, as they have developed more trust of the processes going on within themselves, and have dared to feel their own feelings, live by values which they discover within, and express themselves in their own unique ways.

The General Direction

Let me see if I can state more concisely what is involved in this pattern of movement which I see in clients, the elements of which I have been trying to describe. It seems to mean that the individual moves toward *being,* knowingly and acceptingly, the process which he inwardly and actually *is.* He moves away from being what he is not, from being a façade. He is not trying to be more than he is, with the attendant feelings of insecurity or bombastic defensiveness. He is not trying to be less than he is, with the attendant feelings of guilt or self-depreciation. He is increasingly listening to the deepest recesses of his physiological and emotional being, and finds himself increasingly willing to be, with greater accuracy and depth, that self which he most truly is. One client, as he begins to sense the direction he is taking, asks himself wonderingly and with incredulity in one interview, "You mean if I'd really be what I feel like being, that that would be all right?" His own further experience, and that of many another client, tends toward an affirmative answer. To be what he truly is, this is the path of life which he appears to value most highly, when he is free to move in any direction. It is not simply an intellectual value choice, but seems to be the best description of the groping, tentative, uncertain behaviors by which he moves exploringly toward what he wants to be.

SOME MISAPPREHENSIONS

To many people, the path of life I have been endeavoring to describe seems like a most unsatisfactory path indeed. To the degree that this involves a real difference in values, I simply respect it as a difference. But I have found that sometimes such an attitude is due to certain misapprehensions. In so far as I can I would like to clear these away.

DOES IT IMPLY FIXITY?

To some it appears that to be what one is, is to remain static. They see such a purpose or value as synonymous with being fixed or unchanging. Nothing could be further from the truth. To be what one is, is to enter fully into being a process. Change is facilitated, probably maximized, when one is willing to be what he truly is. Indeed it is the person who is denying his feelings and his reactions who is the person who tends to come for therapy. He has, often for years, been trying to change, but finds himself fixed in these behaviors which he dislikes. It is only as he can become more of himself, can be more of what he has denied in himself, that there is any prospect of change.

Does It Imply Being Evil?

An even more common reaction to the path of life I have been describing is that to be what one truly is would mean to be bad, evil, uncontrolled, destructive. It would mean to unleash some kind of a monster on the world. This is a view which is very well known to me, since I meet it in almost every client. "If I dare to let the feelings flow which are dammed up within me, if by some chance I should live in those feelings, then this would be catastrophe." This is the attitude, spoken or unspoken, of nearly every client as he moves into the experiencing of the unknown aspects of himself. But the whole course of his experience in therapy contradicts these fears. He finds that gradually he can be his anger, when anger is his real reaction, but that such accepted or transparent anger is not destructive. He finds that he can be his fear, but that knowingly to be his fear does not dissolve him. He finds that he can be self-pitying, and it is not "bad." He can feel and be his sexual feelings, or his "lazy" feelings, or his hostile feelings, and the roof of the world does not fall in. The reason seems to be that the more he is able to permit these feelings to flow and to be in him, the more they take their appropriate place in a total harmony of his feelings. He discovers that he has other feelings with which these mingle and find a balance. He feels loving and tender and considerate and cooperative, as well as hostile or lustful or angry. He feels interest and zest and curiosity, as well as laziness or apathy. He feels courageous and venturesome, as well as fearful. His feelings, when he lives closely and acceptingly with their complexity, operate in a constructive harmony rather than sweeping him into some uncontrollably evil path.

Sometimes people express this concern by saying that if an individual were to be what he truly is, he would be releasing the beast in himself. I feel somewhat amused by this, because I think we might take a closer look at the beasts. The lion is often a symbol of the "ravening beast." But what about him? Unless he has been very much warped by contact with humans, he has a number of the qualities I have been describing. To be sure, he kills when he is hungry, but he does not go on a wild rampage of killing, nor does he overfeed himself. He keeps his handsome figure better than some of us. He is helpless and dependent in his puppyhood, but he moves from that to independence. He does not cling to dependence. He is selfish and self-centered in infancy, but in adulthood he shows a reasonable degree of cooperativeness, and feeds, cares for, and protects his young. He satisfies his sexual desires, but this does not mean that he goes on wild and lustful orgies. His various tendencies and urges have a harmony within him. He is, in some basic sense, a constructive and trustworthy member of the species *felis leo*. And what I am trying to suggest is that when one is truly and deeply a unique member of the human species, this is not something which should excite horror. It means instead that one lives fully and openly the complex process of being one of the most widely sensitive, responsive, and creative creatures on this planet. Fully to be one's own uniqueness as a human being, is not, in my experience, a process which would be labeled bad. More appropriate words might be that it is a positive, or a constructive, or a realistic, or a trustworthy process.

Social Implications

Let me turn for a moment to some of the social implications of the path of life I have attempted to describe. I have presented it as a direction which seems to have great meaning for individuals. Does it have, could it have, any meaning or significance for groups or organizations? Would it be a direction which might usefully be chosen by a labor union, a church group, an industrial corporation, a university, a nation? To me it seems that this might be possible. Let us take a look, for example, at the conduct of our own country in its foreign affairs. By and large we find, if we listen to the statements of our leaders during the past several years, and read their documents, that our diplomacy is always based upon high moral purposes; that it is always consistent with the policies we have followed previously; that it involves no selfish desires; and that it has never been mistaken in its judgments and choices. I think perhaps you will agree with me that if we heard an individual speaking in these terms we would recognize at once that this must be a façade, that such statements could not possibly represent the real process going on within himself.

Suppose we speculate for a moment as to how we, as a nation, might present ourselves in our foreign diplomacy if we were openly, knowingly, and acceptingly being what we truly are. I do not know precisely what we are, but I suspect that if we were trying to express ourselves as we are, then our communications with foreign countries would contain elements of this sort.

We as a nation are slowly realizing our enormous strength, and the power and responsibility which go with that strength.

We are moving, somewhat ignorantly and clumsily, toward accepting a position of responsible world leadership.

We make many mistakes. We are often inconsistent.

We are far from perfect.

We are deeply frightened by the strength of Communism, a view of life different from our own.

We feel extremely competitive toward Communism, and we are angry and humiliated when the Russians surpass us in any field.

We have some very selfish foreign interests, such as in the oil in the Middle East.

On the other hand, we have no desire to hold dominion over peoples.

We have complex and contradictory feelings toward the freedom and independence and self-determination of individuals and countries: we desire these and are proud of the past support we have given to such tendencies, and yet we are often frightened by what they may mean.

We tend to value and respect the dignity and worth of each individual, yet when we are frightened, we move away from this direction.

Suppose we presented ourselves in some such fashion, openly and transparently, in our foreign relations. We would be attempting to be the nation which we truly are, in all our complexity and even contradictoriness. What would be the results? To me the results would be similar to the experiences of a client when he is more truly that which he is. Let us look at some of the probable outcomes.

We would be much more comfortable, because we would have nothing to hide.

We could focus on the problem at hand, rather than spending our energies to prove that we are moral or consistent.

We could use all of our creative imagination in solving the problem, rather than in defending ourselves.

We could openly advance both our selfish interests, and our sympathetic concern for others, and let these conflicting desires find the balance which is acceptable to us as a people.

We could freely change and grow in our leadership position, because we would not be bound by rigid concepts of what we have been, must be, ought to be.

We would find that we were much less feared, because others would be less inclined to suspect what lies behind the façade.

We would, by our own openness, tend to bring forth openness and realism on the part of others.

We would tend to work out the solutions of world problems on the basis of the real issues involved, rather than in terms of the façades being worn by the negotiating parties.

In short what I am suggesting by this fantasied example is that nations and organizations might discover, as have individuals, that it is a richly rewarding experience to be what one deeply is. I am suggesting that this view contains the seeds of a philosophical approach to all of life, that it is more than a trend observed in the experience of clients.

SUMMARY

I began this talk with the question each individual asks of himself — what is the goal, the purpose, of my life? I have tried to tell you what I have learned from my clients, who in the therapeutic relationship, with its freedom from threat and freedom of choice, exemplify in their lives a commonality of direction and goal.

I have pointed out that they tend to move away from self-concealment, away from being the expectations of others. The characteristic movement, I have said, is for the client to permit himself freely to be the changing, fluid, process which he is. He moves also toward a friendly openness to what is going on within him — learning to listen sensitively to himself. This means that he is increasingly a harmony of complex sensings and reactions, rather than being the clarity and simplicity of rigidity. It means that as he moves toward acceptance of the "is-ness" of himself, he accepts others increasingly in the same listening, understanding way. He trusts and values the complex inner processes of himself, as they emerge toward expression. He is creatively realistic, and realistically creative. He finds that to be this process in himself is to maximize the rate of change and growth in himself. He is continually engaged in discovering that to be all of himself in this fluid sense is not synonymous with being evil or uncontrolled. It is instead to feel a growing pride in being a sensitive, open, realistic, inner-directed member of the human species, adapting with courage and imagination to the complexities of the changing situation. It means taking continual steps toward being, in awareness and in expression, that which is congruent with one's total organismic reactions. To use Kierkegaard's more aesthetically satisfying terms, it means "to be that self which one truly is." I trust I have made it evident that this is not an easy direction to move, nor one which is ever completed. It is a continuing way of life.

In trying to explore the limits of such a concept, I have suggested that this direction is not a way which is necessarily limited to clients in therapy, nor to individuals seeking to find a purpose in life. It would seem to make the same kind of sense for a group, an organization, or a nation, and would seem to have the same kind of rewarding concomitants.

I recognize quite clearly that this pathway of life which I have outlined is a value choice which is decidedly at variance with the goals usually chosen or behaviorally followed. Yet because it springs from individuals who have more than the usual freedom to choose, and because it seems to express a unified trend in these individuals, I offer it to you for your consideration.

REFERENCES

1. Jacob, P. E. *Changing Values in College.* New Haven: Hazen Foundation, 1956.

2. Kierkegaard, S. *Concluding Unscientific Postscript.* Princeton University Press, 1941.

3. Kierkegaard, S. *The Sickness Unto Death.* Princeton University Press, 1941.

4. Maslow, A. H. *Motivation and Personality.* Harper and Bros., 1954.

5. Morris, C. W. *Varieties of Human Value.* University of Chicago Press, 1956.

6. Seeman, Julius. *The Case of Jim.* Nashville, Tennessee: Educational Testing Bureau, 1957.

7. Whyte, W. H., Jr. *The Organization Man.* Simon & Schuster, 1956.

9

A Therapist's View of the Good Life: The Fully Functioning Person

About 1952 or 1953 I wrote, during one of my winter escapes to warmer climes, a paper I entitled "The Concept of the Fully Functioning Person." It was an attempt to spell out the picture of the person who would emerge if therapy were maximally successful. I was somewhat frightened by the fluid, relativistic, individualistic person who seemed to be the logical outcome of the processes of therapy. I felt two questions. Was my logic correct? If correct, was this the sort of person I valued? To give myself opportunity to mull over these ideas, I had the paper duplicated, and in the ensuing years have distributed hundreds of copies to interested inquirers. As I became more sure of the ideas it contained, I submitted it to one of the major psychological journals. The editor wrote that he would *publish it, but felt that it needed to be cast in a much more conventional psychological framework. He suggested many fundamental changes. This made me feel that it was probably not acceptable to psychologists in the form in which I had written it, and I dropped the idea of publication. Since then it has continued to be a focus of interest for a wide diversity of people, and Dr. Hayakawa has written an article about the concept in the journal of the semanticists,* ETC. *Consequently this was one of the papers which came first to my mind when I contemplated the present book.*

When I re-read it however I found that in the intervening years many of its most central themes and ideas had been absorbed, and perhaps better expressed, in other papers I have included. So, with some reluctance I have again put it aside, and present here instead a paper on my view of the good life, a paper which was based upon "The Fully Functioning Person," and which expresses, I believe, the essential aspects of that paper in briefer and more readable form. My only concession to the past is to give the chapter heading a subtitle.

MY VIEWS regarding the meaning of the good life are largely based upon my experience in working with people in the very close and intimate relationship which is called psychotherapy. These views thus have an empirical or experiential foundation, as contrasted perhaps with a scholarly or philosophical foundation. I have learned what the good life seems to be by observing and participating in the struggle of disturbed and troubled people to achieve that life.

I should make it clear from the outset that this experience I have gained comes from the vantage point of a particular orientation to psychotherapy which has developed over the years. Quite possibly all psychotherapy is basically similar, but since I am less sure of that than I once was, I wish to make it clear that my therapeutic experience has been along the lines that seem to me most effective, the type of therapy termed "client-centered."

Let me attempt to give a very brief description of what this therapy would be like if it were in every respect optimal, since I feel I have learned most about the good life from therapeutic experiences in which a great deal of movement occurred. If the therapy were optimal, intensive as well as extensive, then it would mean that the therapist has been able to enter into an intensely personal and subjective relationship with the client — relating not as a scientist to an object of study, not as a physician expecting to diagnose and cure, but as a person to a person. It would mean that the therapist feels this client to be a person of unconditional self-worth: of value no matter what his condition, his behavior, or his feelings. It would mean that the therapist is genuine, hiding behind no defensive façade, but meeting the client with the feelings which organically he is experiencing. It would mean that the therapist is able to let himself go in understanding this client; that no inner barriers keep him from sensing what it feels like to be the client at each moment of the relationship; and that he can convey something of his empathic understanding to the client. It means that the therapist has been comfortable in entering this relationship fully, without knowing cognitively where it will lead, satisfied with providing a climate which will permit the client the utmost freedom to become himself.

For the client, this optimal therapy would mean an exploration of increasingly strange and unknown and dangerous feelings in himself, the exploration proving possible only because he is gradually realizing that he is accepted unconditionally. Thus he becomes acquainted with elements of his experience which have in the past been denied to awareness as too threatening, too damaging to the structure of the self. He finds himself experiencing these feelings fully, completely, in the relationship, so that for the moment he *is* his fear, or his anger, or his tenderness, or his strength. And as he lives these widely varied feelings, in all their degrees of intensity, he discovers that he has experienced *himself*, that he *is* all these feelings. He finds his behavior changing in constructive fashion in accordance with his newly experienced self. He approaches the realization that he no longer needs to fear what experience may hold, but can welcome it freely as a part of his changing and developing self.

This is a thumbnail sketch of what client-centered therapy comes close to, when it is at its optimum. I give it here simply as a brief picture of the context in which I have formed my views of the good life.

A Negative Observation

As I have tried to live understandingly in the experiences of my clients, I have gradually come to one negative conclusion about the good life. It seems to me that the good life is not any fixed state.

It is not, in my estimation, a state of virtue, or contentment, or nirvana, or happiness. It is not a condition in which the individual is adjusted, or fulfilled, or actualized. To use psychological terms, it is not a state of drive-reduction, or tension-reduction, or homeostasis.

I believe that all of these terms have been used in ways which imply that if one or several of these states is achieved, then the goal of life has been achieved. Certainly, for many people happiness, or adjustment, are seen as states of being which are synonymous with the good life. And social scientists have frequently spoken of the reduction of tension, or the achievement of homeostasis or equilibrium as if these states constituted the goal of the process of living.

So it is with a certain amount of surprise and concern that I realize that my experience supports none of these definitions. If I focus on the experience of those individuals who seem to have evidenced the greatest degree of movement during the therapeutic relationship, and who, in the years following this relationship, appear to have made and to be making real progress toward the good life, then it seems to me that they are not adequately described at all by any of these terms which refer to fixed states of being. I believe they would consider themselves insulted if they were described as "adjusted," and they would regard it as false if they were described as "happy" or "contented," or even "actualized." As I have known them I would regard it as most inaccurate to say that all their drive tensions have been reduced, or that they are in a state of homeostasis. So I am forced to ask myself whether there is any way in which I can generalize about their situation, any definition which I can give of the good life which would seem to fit the facts as I have observed them. I find this not at all easy, and what follows is stated very tentatively.

A Positive Observation

If I attempt to capture in a few words what seems to me to be true of these people, I believe it will come out something like this:

The good life is a *process*, not a state of being.

It is a direction, not a destination.

The direction which constitutes the good life is that which is

selected by the total organism, when there is psychological freedom to move in *any* direction.

This organismically selected direction seems to have certain discernible general qualities which appear to be the same in a wide variety of unique individuals.

So I can integrate these statements into a definition which can at least serve as a basis for consideration and discussion. The good life, from the point of view of my experience, is the process of movement in a direction which the human organism selects when it is inwardly free to move in any direction, and the general qualities of this selected direction appear to have a certain universality.

The Characteristics of the Process

Let me now try to specify what appear to be the characteristic qualities of this process of movement, as they crop up in person after person in therapy.

An Increasing Openness to Experience

In the first place, the process seems to involve an increasing openness to experience. This phrase has come to have more and more meaning for me. It is the polar opposite of defensiveness. Defensiveness I have described in the past as being the organism's response to experiences which are perceived or anticipated as threatening, as incongruent with the individual's existing picture of himself, or of himself in relationship to the world. These threatening experiences are temporarily rendered harmless by being distorted in awareness, or being denied to awareness. I quite literally cannot see, with accuracy, those experiences, feelings, reactions in myself which are significantly at variance with the picture of myself which I already possess. A large part of the process of therapy is the continuing discovery by the client that he is experiencing feelings and attitudes which heretofore he has not been able to be aware of, which he has not been able to "own" as being a part of himself.

If a person could be fully open to his experience, however, every stimulus — whether originating within the organism or in the en-

vironment — would be freely relayed through the nervous system without being distorted by any defensive mechanism. There would be no need of the mechanism of "subception" whereby the organism is forewarned of any experience threatening to the self. On the contrary, whether the stimulus was the impact of a configuration of form, color, or sound in the environment on the sensory nerves, or a memory trace from the past, or a visceral sensation of fear or pleasure or disgust, the person would be "living" it, would have it completely available to awareness.

Thus, one aspect of this process which I am naming "the good life" appears to be a movement away from the pole of defensiveness toward the pole of openness to experience. The individual is becoming more able to listen to himself, to experience what is going on within himself. He is more open to his feelings of fear and discouragement and pain. He is also more open to his feelings of courage, and tenderness, and awe. He is free to live his feelings subjectively, as they exist in him, and also free to be aware of these feelings. He is more able fully to live the experiences of his organism rather than shutting them out of awareness.

INCREASINGLY EXISTENTIAL LIVING

A second characteristic of the process which for me is the good life, is that it involves an increasing tendency to live fully in each moment. This is a thought which can easily be misunderstood, and which is perhaps somewhat vague in my own thinking. Let me try to explain what I mean.

I believe it would be evident that for the person who was fully open to his new experience, completely without defensiveness, each moment would be new. The complex configuration of inner and outer stimuli which exists in this moment has never existed before in just this fashion. Consequently such a person would realize that "What I will be in the next moment, and what I will do, grows out of that moment, and cannot be predicted in advance either by me or by others." Not infrequently we find clients expressing exactly this sort of feeling.

One way of expressing the fluidity which is present in such existential living is to say that the self and personality emerge *from*

experience, rather than experience being translated or twisted to fit preconceived self-structure. It means that one becomes a participant in and an observer of the ongoing process of organismic experience, rather than being in control of it.

Such living in the moment means an absence of rigidity, of tight organization, of the imposition of structure on experience. It means instead a maximum of adaptability, a discovery of structure *in* experience, a flowing, changing organization of self and personality.

It is this tendency toward existential living which appears to me very evident in people who are involved in the process of the good life. One might almost say that it is the most essential quality of it. It involves discovering the structure of experience in the process of living the experience. Most of us, on the other hand, bring a preformed structure and evaluation to our experience and never relinquish it, but cram and twist the experience to fit our preconceptions, annoyed at the fluid qualities which make it so unruly in fitting our carefully constructed pigeonholes. To open one's spirit to what is going on *now*, and to discover in that present process whatever structure it appears to have — this to me is one of the qualities of the good life, the mature life, as I see clients approach it.

An Increasing Trust in His Organism

Still another characteristic of the person who is living the process of the good life appears to be an increasing trust in his organism as a means of arriving at the most satisfying behavior in each existential situation. Again let me try to explain what I mean.

In choosing what course of action to take in any situation, many people rely upon guiding principles, upon a code of action laid down by some group or institution, upon the judgment of others (from wife and friends to Emily Post), or upon the way they have behaved in some similar past situation. Yet as I observe the clients whose experiences in living have taught me so much, I find that increasingly such individuals are able to trust their total organismic reaction to a new situation because they discover to an ever-increasing degree that if they are open to their experience, doing what "feels right" proves to be a competent and trustworthy guide to behavior which is truly satisfying.

As I try to understand the reason for this, I find myself following this line of thought. The person who is fully open to his experience would have access to all of the available data in the situation, on which to base his behavior; the social demands, his own complex and possibly conflicting needs, his memories of similar situations, his perception of the uniqueness of this situation, etc., etc. The data would be very complex indeed. But he could permit his total organism, his consciousness participating, to consider each stimulus, need, and demand, its relative intensity and importance, and out of this complex weighing and balancing, discover that course of action which would come closest to satisfying all his needs in the situation. An analogy which might come close to a description would be to compare this person to a giant electronic computing machine. Since he is open to his experience, all of the data from his sense impressions, from his memory, from previous learning, from his visceral and internal states, is fed into the machine. The machine takes all of these multitudinous pulls and forces which are fed in as data, and quickly computes the course of action which would be the most economical vector of need satisfaction in this existential situation. This is the behavior of our hypothetical person.

The defects which in most of us make this process untrustworthy are the inclusion of information which does *not* belong to this present situation, or the exclusion of information which *does*. It is when memories and previous learnings are fed into the computations as if they were *this* reality, and not memories and learnings, that erroneous behavioral answers arise. Or when certain threatening experiences are inhibited from awareness, and hence are withheld from the computation or fed into it in distorted form, this too produces error. But our hypothetical person would find his organism thoroughly trustworthy, because all of the available data would be used, and it would be present in accurate rather than distorted form. Hence his behavior would come as close as possible to satisfying all his needs — for enhancement, for affiliation with others, and the like.

In this weighing, balancing, and computation, his organism would not by any means be infallible. It would always give the best possible answer for the available data, but sometimes data would be missing. Because of the element of openness to experience, however, any errors, any following of behavior which was not satisfying, would be quickly corrected. The computations, as it were, would always be in process of being corrected, because they would be continually checked in behavior.

Perhaps you will not like my analogy of an electronic computing machine. Let me return to the clients I know. As they become more open to all of their experiences, they find it increasingly possible to trust their reactions. If they "feel like" expressing anger they do so and find that this comes out satisfactorily, because they are equally alive to all of their other desires for affection, affiliation, and relationship. They are surprised at their own intuitive skill in finding behavioral solutions to complex and troubling human relationships. It is only afterward that they realize how surprisingly trustworthy their inner reactions have been in bringing about satisfactory behavior.

The Process of Functioning More Fully

I should like to draw together these three threads describing the process of the good life into a more coherent picture. It appears that the person who is psychologically free moves in the direction of becoming a more fully functioning person. He is more able to live fully in and with each and all of his feelings and reactions. He makes increasing use of all his organic equipment to sense, as accurately as possible, the existential situation within and without. He makes use of all of the information his nervous system can thus supply, using it in awareness, but recognizing that his total organism may be, and often is, wiser than his awareness. He is more able to permit his total organism to function freely in all its complexity in selecting, from the multitude of possibilities, that behavior which in this moment of time will be most generally and genuinely satisfying. He is able to put more trust in his organism in this functioning, not because it is infallible, but because he can be fully open to the consequences of each of his actions and correct them if they prove to be less than satisfying.

He is more able to experience all of his feelings, and is less afraid of any of his feelings; he is his own sifter of evidence, and is more open to evidence from all sources; he is completely engaged in the process of being and becoming himself, and thus discovers that he is soundly and realistically social; he lives more completely in this moment, but learns that this is the soundest living for all time. He is becoming a more fully functioning organism, and because of the awareness of himself which flows freely in and through his experience, he is becoming a more fully functioning person.

SOME IMPLICATIONS

Any view of what constitutes the good life carries with it many implications, and the view I have presented is no exception. I hope that these implications may be food for thought. There are two or three of these about which I would like to comment.

A NEW PERSPECTIVE ON FREEDOM VS DETERMINISM

The first of these implications may not immediately be evident. It has to do with the age-old issue of "free will." Let me endeavor to spell out the way in which this issue now appears to me in a new light.

For some time I have been perplexed over the living paradox which exists in psychotherapy between freedom and determinism. In the therapeutic relationship some of the most compelling subjective experiences are those in which the client feels within himself the power of naked choice. He is *free* — to become himself or to hide behind a façade; to move forward or to retrogress; to behave in ways which are destructive of self and others, or in ways which are enhancing; quite literally free to live or die, in both the physiological and psychological meaning of those terms. Yet as we enter this field of psychotherapy with objective research methods, we are, like any other scientist, committed to a complete determinism. From this point of view every thought, feeling, and action of the client is determined by what preceded it. There can be no such thing as freedom. The dilemma I am trying to describe is no different than that found in other fields — it is simply brought to sharper focus, and appears more insoluble.

This dilemma can be seen in a fresh perspective, however, when we consider it in terms of the definition I have given of the fully functioning person. We could say that in the optimum of therapy the person rightfully experiences the most complete and absolute freedom. He wills or chooses to follow the course of action which is the most economical vector in relationship to all the internal and external stimuli, because it is that behavior which will be most deeply satisfying. But this is the same course of action which from another vantage point may be said to be determined by all the factors in the existential situation. Let us contrast this with the picture of the person who is defensively organized. He wills or chooses to follow a given course of action, but finds that he *cannot* behave in the fashion that he chooses. He is determined by the factors in the existential situation, but these factors include his defensiveness, his denial or distortion of some of the relevant data. Hence it is certain that his behavior will be less than fully satisfying. His behavior is determined, but he is not free to make an effective choice. The fully functioning person, on the other hand, not only experiences, but utilizes, the most absolute freedom when he spontaneously, freely, and voluntarily chooses and wills that which is also absolutely determined.

I am not so naive as to suppose that this fully resolves the issue between subjective and objective, between freedom and necessity. Nevertheless it has meaning for me that the more the person is living the good life, the more he will experience a freedom of choice, and the more his choices will be effectively implemented in his behavior.

Creativity as an Element of the Good Life

I believe it will be clear that a person who is involved in the directional process which I have termed "the good life" is a creative person. With his sensitive openness to his world, his trust of his own ability to form new relationships with his environment, he would be the type of person from whom creative products and creative living emerge. He would not necessarily be "adjusted" to his culture, and he would almost certainly not be a conformist. But at any time and in any culture he would live constructively, in as much harmony with his culture as a balanced satisfaction of needs demanded. In some cultural situations he might in some ways be very unhappy, but he would continue to move toward becoming himself, and to behave in such a way as to provide the maximum satisfaction of his deepest needs.

Such a person would, I believe, be recognized by the student of evolution as the type most likely to adapt and survive under changing environmental conditions. He would be able creatively to make sound adjustments to new as well as old conditions. He would be a fit vanguard of human evolution.

BASIC TRUSTWORTHINESS OF HUMAN NATURE

It will be evident that another implication of the view I have been presenting is that the basic nature of the human being, when functioning freely, is constructive and trustworthy. For me this is an inescapable conclusion from a quarter-century of experience in psychotherapy. When we are able to free the individual from defensiveness, so that he is open to the wide range of his own needs, as well as the wide range of environmental and social demands, his reactions may be trusted to be positive, forward-moving, constructive. We do not need to ask who will socialize him, for one of his own deepest needs is for affiliation and communication with others. As he becomes more fully himself, he will become more realistically socialized. We do not need to ask who will control his aggressive impulses; for as he becomes more open to all of his impulses, his need to be liked by others and his tendency to give affection will be as strong as his impulses to strike out or to seize for himself. He will be aggressive in situations in which aggression is realistically appropriate, but there will be no runaway need for aggression. His total behavior, in these and other areas, as he moves toward being open to all his experience, will be more balanced and realistic, behavior which is appropriate to the survival and enhancement of a highly social animal.

I have little sympathy with the rather prevalent concept that man is basically irrational, and that his impulses, if not controlled, will lead to destruction of others and self. Man's behavior is exquisitely rational, moving with subtle and ordered complexity toward the goals his organism is endeavoring to achieve. The tragedy for most of us is that our defenses keep us from being aware of this rationality, so that consciously we are moving in one direction, while organismically we are moving in another. But in our person who is living the process of the good life, there would be a decreasing number of such barriers, and he would be increasingly a participant in the rationality of his organism. The only control of impulses which would exist, or which would prove necessary, is the natural and internal balancing of one need against another, and the discovery of behaviors which follow the vector most closely approximating the satisfaction of all needs. The experience of extreme satisfaction of one need (for aggression, or sex, etc.) in such a way as to do violence to the satisfaction of other needs (for companionship, tender relationship, etc.) — an experience very common in the defensively organized person — would be greatly decreased. He would participate in the vastly complex self-regulatory activities of his organism — the psychological as well as physiological thermostatic controls — in such a fashion as to live in increasing harmony with himself and with others.

The Greater Richness of Life

One last implication I should like to mention is that this process of living in the good life involves a wider range, a greater richness, than the constricted living in which most of us find ourselves. To be a part of this process means that one is involved in the frequently frightening and frequently satisfying experience of a more sensitive living, with greater range, greater variety, greater richness. It seems to me that clients who have moved significantly in therapy live more intimately with their feelings of pain, but also more vividly with their feelings of ecstasy; that anger is more clearly felt, but so also is love; that fear is an experience they know more deeply, but so is courage. And the reason they can thus live fully in a wider range is that they have this underlying confidence in themselves as trustworthy instruments for encountering life.

I believe it will have become evident why, for me, adjectives such as happy, contented, blissful, enjoyable, do not seem quite appropriate to any general description of this process I have called the good life, even though the person in this process would experience each one of these feelings at appropriate times. But the adjectives which seem more generally fitting are adjectives such as enriching, exciting, rewarding, challenging, meaningful. This process of the good life is not, I am convinced, a life for the faint-hearted. It involves the stretching and growing of becoming more and more of one's potentialities. It involves the courage to be. It means launching oneself fully into the stream of life. Yet the deeply exciting thing about human beings is that when the individual is inwardly free, he chooses as the good life this process of becoming.

PART V

Getting at the Facts: The Place of Research in Psychotherapy

I have endeavored to check
my clinical experience with reality,
but not without some philosophical puzzlement
as to which "reality" is most valid.

10

Persons or Science? A Philosophical Question

This paper stands out for me as one which I found very satisfying to write, and which has continued to be a satisfying expression of my views. I believe that one of the reasons I have liked it is that it was written solely for myself. I had no thought of publishing it or using it for any purpose other than to clarify a growing puzzlement and conflict within myself.

As I look back on it I can recognize the origin of the conflict. It was between the logical positivism in which I was educated, for which I had a deep respect, and the subjectively oriented existential thinking which was taking root in me because it seemed to fit so well with my therapeutic experience.

I am not a student of existential philosophy. I first became acquainted with the work of Søren Kierkegaard and that of Martin Buber at the insistence of some of the theological students at Chicago who were taking work with me. They were sure that I would find the thinking of these men congenial, and in this they were largely correct. While there is much in Kierkegaard, for example, to which I respond not at all, there are, every now and then, deep insights and convictions which beautifully express views I have held but never been able to formulate. Though Kierkegaard lived one hundred years ago, I cannot help but regard him as a sensitive and highly per-

ceptive friend. I think this paper shows my indebtedness to him, mostly in the fact that reading his work loosened me up and made me more willing to trust and express my own experience.

Another helpful element in writing the paper was that I was far away from colleagues, wintering in Taxco, when I wrote the major portion of it. A year later, on the Caribbean island of Grenada, I completed the paper by writing the final section.

As with several of the other papers in this volume, I had it duplicated for reading by my colleagues and students. After several years, at the suggestion of others, I submitted it for publication and it was accepted, rather to my surprise, by the American Psychologist. *I have included it here because it seems to express, better than anything else I have written, the context in which I see research, and makes clear the reason for my "double life" of subjectivity and objectivity.*

INTRODUCTION

THIS IS A HIGHLY PERSONAL DOCUMENT, written primarily for myself, to clarify an issue which has become increasingly puzzling. It will be of interest to others only to the extent that the issue exists for them. I shall therefore describe in this introduction, something of the way in which the paper grew.

As I have acquired experience as a therapist, carrying on the exciting, rewarding experience of psychotherapy, and as I have worked as a scientific investigator to ferret out some of the truth about therapy, I have become increasingly conscious of the gap between these two roles. The better therapist I have become (as I believe I have) the more I have been vaguely aware of my complete subjectivity when I am at my best in this function. And as I have become a better investigator, more "hard-headed" and more scientific (as I believe I have) I have felt an increasing discomfort at the distance between the rigorous objectivity of myself as scientist and the almost mystical subjectivity of myself as therapist. This paper is the result.

What I did first was to let myself go as therapist, and describe, as well as I could do in a brief space, what is the essential nature of psychotherapy as I have lived it with many clients. I would stress the fact that this is a very fluid and personal formulation, and that if it were written by another person, or it were written by me two years ago, or two years hence, it would be different in some respects. Then I let myself go as scientist — as tough-minded fact-finder in this psychological realm, and endeavored to picture the meaning which science can give to therapy. Following this I carried on the debate which existed in me, raising the questions which each point of view legitimately asks the other.

When I had carried my efforts this far I found that I had only sharpened the conflict. The two points of view seemed more than ever irreconcilable. I discussed the material with a seminar of faculty and students, and found their comments very helpful. During the following year I continued to mull over the problem until I began to feel an integration of the two views arising in me. More than a year after the first sections were written I tried to express this tentative and perhaps temporary integration in words.

Thus the reader who cares to follow my struggles in this matter will find that it has quite unconsciously assumed a dramatic form — all of the dramatis personae being contained within myself; First Protagonist, Second Protagonist, The Conflict, and finally, The Resolution. Without more ado let me introduce the first protagonist, myself as therapist, portraying as well as I can, what the *experience* of therapy seems to be.

The Essence of Therapy in Terms of its Experience

I launch myself into the relationship having a hypothesis, or a faith, that my liking, my confidence, and my understanding of the other person's inner world, will lead to a significant process of becoming. I enter the relationship not as a scientist, not as a physician who can accurately diagnose and cure, but as a person, entering into a personal relationship. Insofar as I see him only as an object, the client will tend to become only an object.

I risk myself, because if, as the relationship deepens, what develops is a failure, a regression, a repudiation of me and the relationship by the client, then I sense that I will lose myself, or a part of myself. At times this risk is very real, and is very keenly experienced.

I let myself go into the immediacy of the relationship where it is my total organism which takes over and is sensitive to the relationship, not simply my consciousness. I am not consciously responding in a planful or analytic way, but simply react in an unreflective way to the other individual, my reaction being based, (but not consciously) on my total organismic sensitivity to this other person. I live the relationship on this basis.

The essence of some of the deepest parts of therapy seems to be a unity of experiencing. The client is freely able to experience his feeling in its complete intensity, as a "pure culture," without intellectual inhibitions or cautions, without having it bounded by knowledge of contradictory feelings; and I am able with equal freedom to experience my understanding of this feeling, without any conscious thought about it, without any apprehension or concern as to where this will lead, without any type of diagnostic or analytic thinking, without any cognitive or emotional barriers to a complete "letting go" in understanding. When there is this complete unity, singleness, fullness of experiencing in the relationship, then it acquires the "out-of-this-world" quality which many therapists have remarked upon, a sort of trance-like feeling in the relationship from which both the client and I emerge at the end of the hour, as if from a deep well or tunnel. In these moments there is, to borrow Buber's phrase, a real "I-Thou" relationship, a timeless living in the experience which is *between* the client and me. It is at the opposite pole from seeing the client, or myself, as an object. It is the height of personal subjectivity.

I am often aware of the fact that I do not *know*, cognitively, where this immediate relationship is leading. It is as though both I and the client, often fearfully, let ourselves slip into the stream of becoming, a stream or process which carries us along. It is the fact that the therapist has let himself float in this stream of experience or life previously, and found it rewarding, that makes him each time less fearful of taking the plunge. It is my confidence that makes it easier for the client to embark also, a little bit at a time. It often seems as though this stream of experiencing leads to some goal. Probably the truer statement however, is that its rewarding character lies within the process itself, and that its major reward is that it enables both the client and me, later, independently, to let ourselves go in the process of becoming.

As to the client, as therapy proceeds, he finds that he is daring to become himself, in spite of all the dread consequences which he is sure will befall him if he permits himself to become himself. What does this becoming one's self mean? It appears to mean less fear of the organismic, non-reflective reactions which one has, a gradual growth of trust in and even affection for the complex, varied, rich assortment of feelings and tendencies which exist in one at the organic or organismic level. Consciousness, instead of being the watchman over a dangerous and unpredictable lot of impulses, of which few can be permitted to see the light of day, becomes the comfortable inhabitant of a richly varied society of impulses and feelings and thoughts, which prove to be very satisfactorily self-governing when not fearfully or authoritatively guarded.

Involved in this process of becoming himself is a profound experience of personal choice. He realizes that he can choose to continue to hide behind a façade, or that he can take the risks involved in being himself; that he is a free agent who has it within his power to destroy another, or himself, and also the power to enhance himself and others. Faced with this naked reality of decision, he chooses to move in the direction of being himself.

But being himself doesn't "solve problems." It simply opens up a new way of living in which there is more depth and more height in the experience of his feelings; more breadth and more range. He feels more unique and hence more alone, but he is so much more real that his relationships with others lose their artificial quality, become deeper, more satisfying, and draw more of the realness of the other person into the relationship.

Another way of looking at this process, this relationship, is that it is a learning by the client (and by the therapist, to a lesser extent). But it is a strange type of learning. Almost never is the learning notable by its complexity, and at its deepest the learnings never seem to fit well into verbal symbols. Often the learnings take such simple forms as "I *am* different from others"; "I do feel hatred for him"; "I *am* fearful of feeling dependent"; "I do feel sorry for myself"; "I am self-centered"; "I do have tender and loving feelings"; "I could be what I want to be"; etc. But in spite of their seeming simplicity these learnings are vastly significant in some new way which is very difficult to define. We can think of it in various ways. They are self-appropriated learnings, for one thing, based somehow in experience, not in symbols. They are analogous to the learning of the child who knows that "two and two make four" and who one day playing with two objects and two objects, suddenly realizes in *experience* a totally new learning, that "two and two *do* make four."

Another manner of understanding these learnings is that they are a belated attempt to match symbols with meanings in the world of feelings, an undertaking long since achieved in the cognitive realm. Intellectually, we match carefully the symbol we select with the meaning which an experience has for us. Thus I say something happened "gradually," having quickly (and largely unconsciously) reviewed such terms as "slowly," "imperceptibly," "step-by-step," etc., and rejected them as not carrying the precise shade of meaning of the experience. But in the realm of feelings, we have never learned to attach symbols to experience with any accuracy of meaning. This something which I feel welling up in myself, in the safety of an acceptant relationship — what is it? Is it sadness, is it anger, is it regret, is it sorrow for myself, is it anger at lost opportunities — I stumble around trying out a wide range of symbols, until one "fits," "feels right," seems really to match the organismic experience. In doing this type of thing the client discovers that he has to learn the language of feeling and emotion as if he were an infant learning to speak; often even worse, he finds he must unlearn a false language before learning the true one.

Let us try still one more way of defining this type of learning, this time by describing what it is not. It is a type of learning which cannot be taught. The essence of it is the aspect of self-discovery. With "knowledge" as we are accustomed to think of it, one person can teach it to another, providing each has adequate motivation and ability. But in the significant learning which takes place in therapy, one person *cannot* teach another. The teaching would destroy the learning. Thus I might teach a client that it is safe for him to be himself, that freely to realize his feelings is not dangerous, etc. The more he learned this, the less he would have learned it in the significant, experiential, self-appropriating way. Kierkegaard regards this latter type of learning as true subjectivity, and makes the valid point that there can be no direct communication of it, or even about it. The most that one person can do to further it in another, is to create certain conditions which make this type of learning *possible.* It cannot be compelled.

A final way of trying to describe this learning is that the client gradually learns to symbolize a total and unified state, in which the state of the organism, in experience, feeling, and cognition may all be described in one unified way. To make the matter even more vague and unsatisfactory, it seems quite unnecessary that this symbolization should be expressed. It usually does occur, because the client wishes to communicate at least a portion of himself to the therapist, but it is probably not essential. The only necessary aspect is the inward realization of the total, unified, immediate, "at-this-instant," state of the organism which is me. For example, to realize fully that at this moment the oneness in me is simply that "I am deeply frightened at the possibility of becoming something different" is of the essence of therapy. The client who realizes this will be quite certain to recognize and realize this state of his being when it recurs in somewhat similar form. He will also, in all probability, recognize and realize more fully some of the other existential feelings which occur in him. Thus he will be moving toward a state in which he is more truly himself. He will *be,* in more unified fashion, what he organismically *is,* and this seems to be the essence of therapy.

The Essence of Therapy in Terms of Science

I shall now let the second protagonist, myself as scientist, take over and give his view of this same field.

In approaching the complex phenomena of therapy with the logic and methods of science, the aim is to work toward an *understanding* of the phenomena. In science this means an objective knowledge of events and of functional relationships between events. Science may also give the possibility of increased prediction of and control over these events, but this is not a necessary outcome of scientific endeavor. If the scientific aim were fully achieved in this realm, we would presumably know that, in therapy, certain elements were associated with certain types of outcomes. Knowing this it is likely that we would be able to predict that a particular instance of a therapeutic relationship would have a certain outcome (within certain probability limits) because it involved certain elements. We could then very likely control outcomes of therapy by our manipulation of the elements contained in the therapeutic relationship.

It should be clear that no matter how profound our scientific investigation, we could never by means of it discover any absolute truth, but could only describe relationships which had an increasingly high probability of occurrence. Nor could we ever discover any underlying reality in regard to persons, relationships or the universe. We could only describe relationships between observable events. If science in this field followed the course of science in other fields, the working models of reality which would emerge (in the course of theory building) would be increasingly removed from the reality perceived by the senses. The scientific description of therapy and therapeutic relationships would become increasingly *unlike* these phenomena as they are experienced.

It is evident at the outset that since therapy is a complex phenomenon, measurement will be difficult. Nevertheless "anything that exists can be measured," and since therapy is judged to be a significant relationship, with implications extending far beyond itself, the difficulties may prove to be worth surmounting in order to discover laws of personality and interpersonal relationships.

Since, in client-centered therapy, there already exists a crude theory (though not a theory in the strictly scientific sense) we have a starting point for the selection of hypotheses. For purposes of this discussion, let us take some of the crude hypotheses which can be drawn from this theory, and see what a scientific approach will do with them. We will, for the time being, omit the translation of the total theory into a formal logic which would be acceptable and consider only a few of the hypotheses.

Let us first state three of these in their crude form.

1. Acceptance of the client by the therapist leads to an increased acceptance of self by the client.

2. The more the therapist perceives the client as a person rather than as an object, the more the client will come to perceive himself as a person rather than an object.

3. In the course of therapy an experiential and effective type of learning about self takes place in the client.

How would we go about translating each of these* into operational terms and how would we test the hypotheses? What would be the general outcomes of such testing?

This paper is not the place for a detailed answer to these questions, but research already carried on supplies the answers in a general way. In the case of the first hypothesis, certain devices for measuring acceptance would be selected or devised. These might be attitude tests, objective or projective, Q technique or the like. Presumably the same instruments, with slightly different instructions or mind set, could be used to measure the therapist's acceptance of the client, and the client's acceptance of self. Operationally then, the degree of therapist acceptance would be equated to a certain score on this instrument. Whether client self-acceptance changed during therapy would be indicated by pre- and post-measurements. The relationship of any change to therapy would be determined by comparison of changes in therapy to changes during a control period or in a control group. We would finally be able to say whether a relationship existed between therapist acceptance and client self-acceptance, as operationally defined, and the correlation between the two.

The second and third hypotheses involve real difficulty in meas-

* It may be surprising to some to find hypotheses regarding such subjective experience treated as matters for an objective science. Yet the best thinking in psychology has gone far beyond a primitive behaviorism, and has recognized that the objectivity of psychology as science rests upon its method, not upon its content. Thus the most subjective feelings, apprehensions, tensions, satisfactions, or reactions, may be dealt with scientifically, providing only that they may be given clearcut operational definition. Stephenson, among others, presents this point of view forcefully (in his Postulates of Behaviorism) and through his Q Technique, has contributed importantly to the objectification of such subjective materials for scientific study.

urement, but there is no reason to suppose that they could not be objectively studied, as our sophistication in psychological measurement increases. Some type of attitude test or Q-sort might be the instrument for the second hypothesis, measuring the attitude of therapist toward client, and of client toward self. In this case the continuum would be from objective regard of an external object to a personal and subjective experiencing. The instrumentation for hypothesis three might be physiological, since it seems likely that experiential learning has physiologically measurable concomitants. Another possibility would be to infer experiential learning from its effectiveness, and thus measure the effectiveness of learning in different areas. At the present stage of our methodology hypothesis three might be beyond us, but certainly within the foreseeable future, it too could be given operational definition and tested.

The findings from these studies would be of this order. Let us become suppositious, in order to illustrate more concretely. Suppose we find that therapist acceptance leads to client self-acceptance, and that the correlation is in the neighborhood of .70 between the two variables. In hypothesis two we might find the hypothesis unsupported, but find that the more the therapist regarded the client as a person, the more the client's self-acceptance increased. Thus we would have learned that person-centeredness is an element of acceptance, but that it has little to do with the client becoming more of a person to himself. Let us also suppose hypothesis three upheld with experiential learning of certain describable sorts taking place much more in therapy than in the control subjects.

Glossing over all the qualifications and ramifications which would be present in the findings, and omitting reference to the unexpected leads into personality dynamics which would crop up (since these are hard to imagine in advance) the preceding paragraph gives us some notion of what science can offer in this field. It can give us a more and more exact description of the events of therapy and the changes which take place. It can begin to formulate some tentative laws of the dynamics of human relationships. It can offer public and replicable statements, that if certain operationally definable conditions exist in the therapist or in the relationship, then certain client behaviors may be expected with a known degree of probability. It can presumably do this for the field of therapy and personality change as it is in the process of doing for such fields as perception and learning. Eventually theoretical formulations should draw together these different areas, enunciating the laws which appear to govern alteration in human behavior, whether in the situations we classify as perception, those we classify as learning, or the more global and molar changes which occur in therapy, involving both perception and learning.

Some Issues

Here are two very different methods of perceiving the essential aspects of psychotherapy, two very different approaches to forging ahead into new territory in this field. As presented here, and as they frequently exist, there seems almost no common meeting ground between the two descriptions. Each represents a vigorous way of seeing therapy. Each seems to be an avenue to the significant truths of therapy. When each of these views are held by different individuals or groups, they constitute a basis of sharp disagreement. When each of these approaches seems true to one individual, like myself, then he feels himself conflicted by these two views. Though they may superficially be reconciled, or regarded as complementary to each other, they seem to me to be basically antagonistic in many ways. I should like to raise certain issues which these two viewpoints pose for me.

The Scientist's Questions

First let me pose some of the questions which the scientific viewpoint asks of the experiential (using scientific and experiential simply as loose labels to indicate the two views). The hard-headed scientist listens to the experiential account, and raises several searching questions.

1. First of all he wants to know, "How can you know that this account, or any account given at a previous or later time, is true? How do you know that it has any relationship to reality? If we are to rely on this inner and subjective experience as being the truth

about human relationships or about ways of altering personality, then Yogi, Christian Science, dianetics, and the delusions of a psychotic individual who believes himself to be Jesus Christ, are all true, just as true as this account. Each of them represents the truth as perceived inwardly by some individual or group of individuals. If we are to avoid this morass of multiple and contradictory truths, we must fall back on the only method we know for achieving an ever-closer approximation to reality, the scientific method."

2. "In the second place, this experiential approach shuts one off from improving his therapeutic skill, or discovering the less than satisfactory elements in the relationship. Unless one regards the present description as a perfect one, which is unlikely, or the present level of experience in the therapeutic relationship as being the most effective possible, which is equally unlikely, then there are unknown flaws, imperfections, blind spots, in the account as given. How are these to be discovered and corrected? The experiential approach can offer nothing but a trial and error process for achieving this, a process which is slow and which offers no real guarantee of achieving this goal. Even the criticisms or suggestions of others are of little help, since they do not arise from within the experience and hence do not have the vital authority of the relationship itself. But the scientific method, and the procedures of a modern logical positivism, have much to offer here. Any experience which can be described at all can be described in operational terms. Hypotheses can be formulated and put to test, and the sheep of truth can thus be separated from the goats of error. This seems the only sure road to improvement, self-correction, growth in knowledge."

3. The scientist has another comment to make. "Implicit in your description of the therapeutic experience seems to be the notion that there are elements in it which *cannot* be predicted — that there is some type of spontaneity or (excuse the term) free will operative here. You speak as though some of the client's behavior — and perhaps some of the therapist's — is not caused, is not a link in a sequence of cause and effect. Without desiring to become metaphysical, may I raise the question as to whether this is defeatism? Since surely we can discover what causes *much* of behavior — you yourself speak of creating the conditions where certain behavioral re-

sults follow — then why give up at any point? Why not at least *aim* toward uncovering the causes of *all* behavior? This does not mean that the individual must regard himself as an automaton, but in our search for the facts we shall not be hampered by a belief that some doors are closed to us."

4. Finally, the scientist cannot understand why the therapist, the experientialist, should challenge the one tool and method which is responsible for almost all the advances which we value. "In the curing of disease, in the prevention of infant mortality, in the growing of larger crops, in the preservation of food, in the manufacture of all the things that make life comfortable, from books to nylon, in the understanding of the universe, what is the foundation stone? It is the method of science, applied to each of these, and to many other problems. It is true that it has improved methods of warfare, too, serving man's destructive as well as his constructive purposes, but even here the potentiality for social usefulness is very great. So why should we doubt this same approach in the social science field? To be sure advances here have been slow, and no law as fundamental as the law of gravity has as yet been demonstrated, but are we to give up this approach out of impatience? What possible alternative offers equal hope? If we are agreed that the social problems of the world are very pressing indeed, if psychotherapy offers a window into the most crucial and significant dynamics of change in human behavior, then surely the course of action is to apply to psychotherapy the most rigorous canons of scientific method, on as broad a scale as possible, in order that we may most rapidly approach a tentative knowledge of the laws of individual behavior and of attitudinal change."

The Questions of the Experientialist

While the scientist's questions may seem to some to settle the matter, his comments are far from being entirely satisfying to the therapist who has lived the experience of therapy. Such an individual has several points to make in regard to the scientific view.

1. "In the first place," this "experientialist" points out, "science always has to do with the other, the object. Various logicians of science, including Stevens, the psychologist, show that it is a basic

element of science that it always has to do with the observable object, the observable other. This is true, even if the scientist is experimenting on himself, for to that degree he treats himself as the observable other. It never has anything to do with the experiencing me. Now does not this quality of science mean that it must forever be irrelevant to an experience such as therapy, which is intensely personal, highly subjective in its inwardness, and dependent entirely on the relationship of two individuals each of whom is an experiencing me? Science can of course study the events which occur, but always in a way which is irrelevant to what is occurring. An analogy would be to say that science can conduct an autopsy of the dead events of therapy, but by its very nature it can never enter into the living physiology of therapy. It is for this reason that therapists recognize — usually intuitively — that any advance in therapy, any fresh knowledge of it, any significant new hypotheses in regard to it — must come from the experience of the therapists and clients, and can never come from science. Again to use an analogy. Certain heavenly bodies were discovered solely from examination of the scientific measurements of the courses of the stars. Then the astronomers searched for these hypothesized bodies and found them. It seems decidedly unlikely that there will ever be a similar outcome in therapy, since science has nothing to say about the internal personal experience which 'I' have in therapy. It can only speak of the events which occur in 'him.' "

2. "Because science has as its field the 'other,' the 'object,' it means that everything it touches is transformed into an object. This has never presented a problem in the physical sciences. In the biological sciences it has caused certain difficulties. A number of medical men feel some concern as to whether the increasing tendency to view the human organism as an object, in spite of its scientific efficacy, may not be unfortunate for the patient. They would prefer to see him again regarded as a person. It is in the social sciences, however, that this becomes a genuinely serious issue. It means that the people studied by the social scientist are always objects. In therapy, both client and therapist become objects for dissection, but not persons with whom one enters a living relationship. At first glance, this may not seem important. We may say that only in his role as scien-

tist does the individual regard others as objects. He can also step out of this role and become a person. But if we look a little further we will see that this is a superficial answer. If we project ourselves into the future, and suppose that we had the answers to most of the questions which psychology investigates today, what then? Then we would find ourselves increasingly impelled to treat all others, and even ourselves, as objects. The knowledge of all human relationships would be so great that we would know it rather than live the relationships unreflectively. We see some foretaste of this in the attitude of sophisticated parents who know that affection 'is good for the child.' This knowledge frequently stands in the way of their being themselves, freely, unreflectively — affectionate or not. Thus the development of science in a field like therapy is either irrelevant to the experience, or may actually make it more difficult to live the relationship as a personal, experiential event."

3. The experientialist has a further concern. "When science transforms people into objects, as mentioned above, it has another effect. The end result of science is to lead toward manipulation. This is less true in fields like astronomy, but in the physical and social sciences, the knowledge of the events and their relationships lead to manipulation of some of the elements of the equation. This is unquestionably true in psychology, and would be true in therapy. If we know all about how learning takes place, we use that knowledge to manipulate persons as objects. This statement places no value judgment on manipulation. It may be done in highly ethical fashion. We may even manipulate ourselves as objects, using such knowledge. Thus, knowing that learning takes place more rapidly with repeated review rather than long periods of concentration on one lesson, I may use this knowledge to manipulate my learning in Spanish. But knowledge is power. As I learn the laws of learning I use them to manipulate others through advertisements, through propaganda, through prediction of their responses and the control of those responses. It is not too strong a statement to say that the growth of knowledge in the social sciences contains within itself a powerful tendency toward social control, toward control of the many by the few. An equally strong tendency is toward the weakening or destruction of the existential person. When all are regarded as objects, the sub-

jective individual, the inner self, the person in the process of becoming, the unreflective consciousness of being, the whole inward side of living life, is weakened, devalued, or destroyed. Perhaps this is best exemplified by two books. Skinner's *Walden Two* is a psychologist's picture of paradise. To Skinner it must have seemed desirable, unless he wrote it as a tremendous satire. At any rate it is a paradise of manipulation, in which the extent to which one can be a person is greatly reduced, unless one can be a member of the ruling council. Huxley's *Brave New World* is frankly satire, but portrays vividly the loss of personhood which he sees as associated with increasing psychological and biological knowledge. Thus, to put it bluntly, it seems that a developing social science (as now conceived and pursued) leads to social dictatorship and individual loss of personhood. The dangers perceived by Kierkegaard a century ago in this respect seem much more real now, with the increase of knowledge, than they could have then."

4. "Finally," says the experientialist, "doesn't all this point to the fact that ethics is a more basic consideration than science? I am not blind to the value of science as a tool, and am aware that it can be a very valuable tool. But unless it is the tool of ethical *persons*, with all that the term persons implies, may it not become a Juggernaut? We have been a long time recognizing this issue, because in physical science it took centuries for the ethical issue to become crucial, but it has at last become so. In the social sciences the ethical issues arise much more quickly, because persons are involved. But in psychotherapy the issue arises most quickly and most deeply. Here is the maximizing of all that is subjective, inward, personal; here a relationship is lived, not examined, and a person, not an object, emerges; a person who feels, chooses, believes, acts, not as an automaton, but as a person. And here too is the ultimate in science — the objective exploration of the most subjective aspects of life; the reduction to hypotheses, and eventually to theorems, of all that has been regarded as most personal, most completely inward, most thoroughly a private world. And because these two views come so sharply into focus here, we must make a choice — an ethical personal choice of values. We may do it by default, by not raising the question. We may be able to make a choice which will somehow

conserve both values — but choose we must. And I am asking that we think long and hard before we give up the values that pertain to being a person, to experiencing, to living a relationship, to becoming, that pertain to one's self as a process, to one's self in the existential moment, to the inward subjective self that lives."

The Dilemma

There you have the contrary views as they occur sometimes explicitly, more often implicitly, in current psychological thinking. There you have the debate as it exists in me. Where do we go? What direction do we take? Has the problem been correctly described or is it fallacious? What are the errors of perception? Or if it is essentially as described, must we choose one or the other? And if so, which one? Or is there some broader, more inclusive formulation which can happily encompass both of these views without damage to either?

A Changed View of Science

In the year which has elapsed since the foregoing material was written, I have from time to time discussed the issues with students, colleagues and friends. To some of them I am particularly indebted for ideas which have taken root in me.* Gradually I have come to believe that the most basic error in the original formulation was in the description of science. I should like, in this section, to attempt to correct that error, and in the following section to reconcile the revised points of view.

The major shortcoming was, I believe, in viewing science as something "out there," something spelled with a capital S, a "body of knowledge" existing somewhere in space and time. In common with many psychologists I thought of science as a systematized and or-

* I would like to mention my special debt to discussions with, and published and unpublished papers by Robert M. Lipgar, Ross L. Mooney, David A. Rodgers and Eugene Streich. My own thinking has fed so deeply on theirs, and become so intertwined with theirs, that I would be at a loss to acknowledge specific obligations. I only know that in what follows there is much which springs from them, through me. I have also profited from correspondence regarding the paper with Anne Roe and Walter Smet.

ganized collection of tentatively verified facts, and saw the methodology of science as the socially approved means of accumulating this body of knowledge, and continuing its verification. It has seemed somewhat like a reservoir into which all and sundry may dip their buckets to obtain water — with a guarantee of 99% purity. When viewed in this external and impersonal fashion, it seems not unreasonable to see Science not only as discovering knowledge in lofty fashion, but as involving depersonalization, a tendency to manipulate, a denial of the basic freedom of choice which I have met experientially in therapy. I should like now to view the scientific approach from a different, and I hope, a more accurate perspective.

Science in Persons

Science exists only in people. Each scientific project has its creative inception, its process, and its tentative conclusion, in a person or persons. Knowledge — even scientific knowledge — is that which is subjectively acceptable. Scientific knowledge can be communicated only to those who are subjectively ready to receive its communication. The utilization of science also occurs only through people who are in pursuit of values which have meaning for them. These statements summarize very briefly something of the change in emphasis which I would like to make in my description of science. Let me follow through the various phases of science from this point of view.

The Creative Phase

Science has its inception in a particular person who is pursuing aims, values, purposes, which have personal and subjective meaning for him. As a part of this pursuit, he, in some area, "wants to find out." Consequently, if he is to be a good scientist, he immerses himself in the relevant experience, whether that be the physics laboratory, the world of plant or animal life, the hospital, the psychological laboratory or clinic, or whatever. This immersion is complete and subjective, similar to the immersion of the therapist in therapy, described previously. He senses the field in which he is interested, he lives it. He does more than "think" about it — he lets his organism take over and react to it, both on a knowing and on an unknowing level. He comes to sense more than he could possibly verbalize about his field, and reacts organismically in terms of relationships which are not present in his awareness.

Out of this complete subjective immersion comes a creative forming, a sense of direction, a vague formulation of relationships hitherto unrecognized. Whittled down, sharpened, formulated in clearer terms, this creative forming becomes a hypothesis — a statement of a tentative, personal, subjective faith. The scientist is saying, drawing upon all his known and unknown experience, that "I have a hunch that such and such a relationship exists, and the existence of this phenomenon has relevance to my personal values."

What I am describing is the initial phase of science, probably its most important phase, but one which American scientists, particularly psychologists, have been prone to minimize or ignore. It is not so much that it has been denied as that it has been quickly brushed off. Kenneth Spence has said that this aspect of science is "simply taken for granted."* Like many experiences taken for granted, it also tends to be forgotten. It is indeed in the matrix of immediate personal, subjective experience that all science, and each individual scientific research, has its origin.

Checking with Reality

The scientist has then creatively achieved his hypothesis, his tentative faith. But does it check with reality? Experience has shown each one of us that it is very easy to deceive ourselves, to believe something which later experience shows is not so. How can I tell whether this tentative belief has some real relationship to observed facts? I can use, not one line of evidence only, but several. I can surround my observation of the facts with various precautions to make sure I am not deceiving myself. I can consult with others

* It may be pertinent to quote the sentences from which this phrase is taken. ". . . the data of all sciences have the same origin — namely, the immediate experience of an observing person, the scientist himself. That is to say, immediate experience, the initial matrix out of which all sciences develop, is no longer considered a matter of concern for the scientist qua scientist. He simply takes it for granted and then proceeds to the task of describing the events occurring in it and discovering and formulating the nature of the relationships holding among them." Kenneth W. Spence, in *Psychological Theory*, ed. by M. H. Marx (New York: Macmillan, 1951), p. 173.

who have also been concerned with avoiding self-deception, and learn useful ways of catching myself in unwarranted beliefs, based on misinterpretation of observations. I can, in short, begin to use all the elaborate methodology which science has accumulated. I discover that stating my hypothesis in operational terms will avoid many blind alleys and false conclusions. I learn that control groups can help me to avoid drawing false inferences. I learn that correlations, and t tests and critical ratios and a whole array of statistical procedures can likewise aid me in drawing only reasonable inferences.

Thus scientific methodology is seen for what it truly is — a way of preventing me from deceiving myself in regard to my creatively formed subjective hunches which have developed out of the relationship between me and my material. It is in this context, and perhaps only in this context, that the vast structure of operationism, logical positivism, research design, tests of significance, etc. have their place. They exist, not for themselves, but as servants in the attempt to check the subjective feeling or hunch or hypothesis of a person with the objective fact.

And even throughout the use of such rigorous and impersonal methods, the important choices are all made subjectively by the scientist. To which of a number of hypotheses shall I devote time? What kind of control group is most suitable for avoiding self-deception in this particular research? How far shall I carry the statistical analysis? How much credence may I place in the findings? Each of these is necessarily a subjective personal judgment, emphasizing that the splendid structure of science rests basically upon its subjective use by persons. It is the best instrument we have yet been able to devise to check upon our organismic sensing of the universe.

THE FINDINGS

If, as scientist, I like the way I have gone about my investigation, if I have been open to all the evidence, if I have selected and used intelligently all the precautions against self-deception which I have been able to assimilate from others or to devise myself, then I will give my tentative belief to the findings which have emerged. I will regard them as a springboard for further investigation and further seeking.

It seems to me that in the best of science, the primary purpose is to provide a more satisfactory and dependable hypothesis, belief, faith, for the investigator himself. To the extent that the scientist is endeavoring to prove something to someone else — an error into which I have fallen more than once — then I believe he is using science to bolster a personal insecurity, and is keeping it from its truly creative role in the service of the person.

In regard to the findings of science, the subjective foundation is well shown in the fact that at times the scientist may refuse to believe his own findings. "The experiment showed thus and so, but I believe it is wrong," is a theme which every scientist has experienced at some time or other. Some very fruitful discoveries have grown out of the persistent *disbelief*, by a scientist, in his own findings and those of others. In the last analysis he may place more trust in his total organismic reactions than in the methods of science. There is no doubt that this can result in serious error as well as in scientific discoveries, but it indicates again the leading place of the subjective in the use of science.

Communication of Scientific Findings

Wading along a coral reef in the Caribbean this morning, I saw a large blue fish — I think. If you, quite independently, saw it too, then I feel more confident in my own observation. This is what is known as intersubjective verification, and it plays an important part in our understanding of science. If I take you (whether in conversation or in print or behaviorally) through the steps I have taken in an investigation, and it seems to you too that I have not deceived myself, and that I have indeed come across a new relationship which is relevant to my values, and that I am justified in having a tentative faith in this relationship, then we have the beginnings of Science with a capital S. It is at this point that we are likely to think we have created a body of scientific knowledge. Actually there is no such body of knowledge. There are only tentative beliefs, existing subjectively, in a number of different persons. If these beliefs are not tentative, then what exists is dogma, not science. If on the other hand, no one but the investigator believes the finding then this finding is either a personal and deviant matter, an instance of psychopathology, or else it is an unusual truth discovered by a genius, which as yet no one is subjectively ready to believe. This leads me to comment on the group which can put tentative faith in any given scientific finding.

COMMUNICATION TO WHOM?

It is clear that scientific findings can be communicated only to those who have agreed to the same ground rules of investigation. The Australian bushman will be quite unimpressed with the findings of science regarding bacterial infection. He knows that illness truly is caused by evil spirits. It is only when he too agrees to scientific method as a good means of preventing self-deception, that he will be likely to accept its findings.

But even among those who have adopted the ground rules of science, tentative belief in the findings of a scientific research can only occur where there is a subjective readiness to believe. One could find many examples. Most psychologists are quite ready to believe evidence showing that the lecture system produces significant increments of learning, and quite unready to believe that the turn of an unseen card may be called through an ability labelled extra-sensory perception. Yet the scientific evidence for the latter is considerably more impeccable than for the former. Likewise when the so-called "Iowa studies" first came out, indicating that intelligence might be considerably altered by environmental conditions, there was great disbelief among psychologists, and many attacks on the imperfect scientific methods used. The scientific evidence for this finding is not much better today than it was when the Iowa studies first appeared, but the subjective readiness of psychologists to believe such a finding has altered greatly. A historian of science has noted that empiricists, had they existed at the time, would have been the first to disbelieve the findings of Copernicus.

It appears then that whether I believe the scientific findings of others, or those from my own studies, depends in part on my readiness to put a tentative belief in such findings.* One reason we are

* One example from my own experience may suffice. In 1941 a research study done under my supervision showed that the future adjustment of

not particularly aware of this subjective fact is that in the physical sciences particularly, we have gradually adopted a very large area of experience in which we are ready to believe any finding which can be shown to rest upon the rules of the scientific game, properly played.

The Use of Science

But not only is the origin, process, and conclusion of science something which exists only in the subjective experience of persons — so also is its utilization. "Science" will never depersonalize, or manipulate, or control individuals. It is only persons who can and will do that. That is surely a most obvious and trite observation, yet a deep realization of it has had much meaning for me. It means that the use which will be made of scientific findings in the field of personality is and will be a matter of subjective personal choice — the same type of choice as a person makes in therapy. To the extent that he has defensively closed off areas of his experience from awareness, the person is more likely to make choices which are socially destructive. To the extent that he is open to all phases of his experience we may be sure that this person will be more likely to use the findings and methods of science (or any other tool or capacity) in a manner which is personally and socially constructive.* There is, in actuality then, no threatening entity of "Science" which can in any way affect our destiny. There are only people. While many of them are indeed

---

delinquent adolescents was best predicted by a measure of their realistic self-understanding and self-acceptance. The instrument was a crude one, but it was a better predictor than measures of family environment, hereditary capacities, social milieu, and the like. At that time I was simply not ready to believe such a finding, because my own belief, like that of most psychologists, was that such factors as the emotional climate in the family and the influence of the peer group were the real determinants of future delinquency and non-delinquency. Only gradually, as my experience with psychotherapy continued and deepened, was it possible for me to give my tentative belief to the findings of this study and of a later one (1944) which confirmed it. (For a report of these two studies see "The role of self-understanding in the prediction of behavior" by C. R. Rogers, B. L. Kell, and H. McNeil, *J. Consult. Psychol., 12*, 1948, pp. 174–186.

* I have spelled out more fully the rationale for this view in another paper — "Toward a Theory of Creativitv."

threatening and dangerous in their defensiveness, and modern scientific knowledge multiplies the social threat and danger, this is not the whole picture. There are two other significant facets. (1) There are many persons who are relatively open to their experience and hence likely to be socially constructive. (2) Both the subjective experience of psychotherapy and the scientific findings regarding it indicate that individuals are motivated to change, and may be helped to change, in the direction of greater openness to experience, and hence in the direction of behavior which is enhancing of self and society, rather than destructive.

To put it briefly, Science can never threaten us. Only persons can do that. And while individuals can be vastly destructive with the tools placed in their hands by scientific knowledge, this is only one side of the picture. We already have subjective and objective knowledge of the basic principles by which individuals may achieve the more constructive social behavior which is natural to their organismic process of becoming.

A New Integration

What this line of thought has achieved for me is a fresh integration in which the conflict between the "experientialist" and the "scientist" tends to disappear. This particular integration may not be acceptable to others, but it does have meaning to me. Its major tenets have been largely implicit in the preceding section, but I will try to state them here in a way which takes cognizance of the arguments between the opposing points of view.

Science, as well as therapy, as well as all other aspects of living, is rooted in and based upon the immediate, subjective experience of a person. It springs from the inner, total, organismic experiencing which is only partially and imperfectly communicable. It is one phase of subjective living.

It is because I find value and reward in human relationships that I enter into a relationship known as therapeutic, where feelings and cognition merge into one unitary experience which is lived rather than examined, in which awareness is non-reflective, and where I am participant rather than observer. But because I am curious about the exquisite orderliness which appears to exist in the universe and in this relationship I can abstract myself from the experience and look upon it as an observer, making myself and/or others the objects of that observation. As observer I use all of the hunches which grow out of the living experience. To avoid deceiving myself as observer, to gain a more accurate picture of the order which exists, I make use of all the canons of science. Science is not an impersonal something, but simply a person living subjectively another phase of himself. A deeper understanding of therapy (or of any other problem) may come from living it, or from observing it in accordance with the rules of science, or from the communication within the self between the two types of experience. As to the subjective experience of choice, it is not only primary in therapy, but it is also primary in the use of scientific method by a person.

What I will do with the knowledge gained through scientific method — whether I will use it to understand, enhance, enrich, or use it to control, manipulate and destroy — is a matter of subjective choice dependent upon the values which have personal meaning for me. If, out of fright and defensiveness, I block out from my awareness large areas of experience, — if I can see only those facts which support my present beliefs, and am blind to all others — if I can see only the objective aspects of life, and cannot perceive the subjective — if in any way I cut off my perception from the full range of its actual sensitivity — then I am likely to be socially destructive, whether I use as tool the knowledge and instruments of science, or the power and emotional strength of a subjective relationship. And on the other hand if I am open to my experience, and can permit all of the sensings of my intricate organism to be available to my awareness, then I am likely to use myself, my subjective experience, *and* my scientific knowledge, in ways which are realistically constructive.

This then is the degree of integration I have currently been able to achieve between two approaches first experienced as conflicting. It does not completely resolve all the issues posed in the earlier section, but it seems to point toward a resolution. It rewrites the problem or reperceives the issue, by putting the subjective, existential person, with the values which he holds, at the foundation and the root of the therapeutic relationship and of the scientific relationship. For science too, at its inception, is an "I-Thou" relationship with a person or persons. And only as a subjective person can I enter into either of these relationships.

11

Personality Change in Psychotherapy

The paper which follows gives a few of the salient features of a very large scale research carried on at the University of Chicago Counseling Center from 1950–1954, made possible by the generous support of the Rockefeller Foundation, through its Medical Sciences Division. I was invited to present a paper to the Fifth International Congress on Mental Health in Toronto, in 1954, and chose to attempt to describe certain portions of that program. Within a month of the delivery of this paper, our book describing the whole program was published by the University of Chicago Press. Although Rosalind Dymond and I served as editors as well as authors of certain portions of the book, the other authors deserve equal credit for the book and for the vast amount of work from which this paper skims a few of the more striking points. These other authors are: John M. Butler, Desmond Cartwright, Thomas Gordon, Donald L. Grummon, Gerard V. Haigh, Eve S. John, Esselyn C. Rudikoff, Julius Seeman, Rolland R. Tougas, and Manuel J. Vargas.

A special reason for including this presentation in this volume is that it gives in brief form some of the exciting progress we have made in the measurement of that changing, nebulous, highly significant and determining aspect of personality, the self.

IT IS THE PURPOSE of this paper to present some of the high lights of the experience which I and my colleagues have had as we endeavored to measure, by objective scientific methods, the outcomes of one form of individual psychotherapy. In order to make these high lights understandable, I shall describe briefly the context in which this research undertaking has been carried on.

For many years I have been working, with my psychologist colleagues, in the field of psychotherapy. We have been trying to learn, from our experience in carrying on psychotherapy, what is effective in bringing about constructive change in the personality and behavior of the maladjusted or disturbed person seeking help. Gradually we have formulated an approach to psychotherapy, based upon this experience, which has variously been termed non-directive or client-centered. This approach and its theoretical rationale have been described in a number of books (1, 2, 5, 6, 8) and many articles.

It has been one of our persistent aims to subject the dynamics of therapy and the results of therapy to rigorous research investigation. It is our belief that psychotherapy is a deeply subjective existential experience in both client and therapist, full of complex subtleties, and involving many nuances of personal interaction. Yet it is also our conviction that if this experience is a significant one, in which deep learnings bring about personality change, then such changes should be amenable to research investigation.

Over the past fourteen years we have made many such research studies, of both the process and the outcomes of this form of therapy. (See 5, particularly chapters 2, 4, and 7, for a summarized account of this body of research.) During the past five years, at the Counseling Center of the University of Chicago, we have been pushing forward the boundaries of such research by means of a coordinated series of investigations designed to throw light upon the outcomes of this form of psychotherapy. It is from this current research program that I wish to present certain significant features.

Three Aspects of Our Research

The three aspects of our research which would, I believe, have the greatest amount of meaning to this audience, are these.

1. The criteria which we have used in our study of psychotherapy, criteria which depart from conventional thinking in this area.

2. The design of the research, in which we have solved certain difficulties which have hitherto stood in the way of clear-cut results.

3. The progress we have made in measuring subtle subjective phenomena in an objective fashion.

These three elements in our program could be utilized in any attempt to measure personality change. They are therefore applicable to investigations of any form of psychotherapy, or to the research study of any procedure designed to bring about alteration in personality or behavior.

Let us now turn to these three elements I have mentioned, taking them up in order.

The Criteria for the Research

What is the criterion for research in psychotherapy? This is a most perplexing issue which we faced early in our planning. There is widespread acceptance of the idea that the purpose of research in this field is to measure the degree of "success" in psychotherapy, or the degree of "cure" achieved. While we have not been uninfluenced by such thinking, we have, after careful consideration, given up these concepts because they are undefinable, are essentially value judgments, and hence cannot be a part of the science of this field. There is no general agreement as to what constitutes "success" — whether it is removal of symptoms, resolution of conflicts, improvement in social behavior, or some other type of change. The concept of "cure" is entirely inappropriate, since in most of these disorders we are dealing with learned behavior, not with a disease.

As a consequence of our thinking, we have not asked in our research, "Was success achieved? Was the condition cured?" Instead we have asked a question which is scientifically much more defensible, namely, "What are the concomitants of therapy?"

In order to have a basis for answering this question we have taken the theory of psychotherapy which we have been developing and have drawn from it the theoretical description of those changes which we hypothesized as occurring in therapy. The purpose of the research is to determine whether the changes which are hypothesized do or do not occur in measurable degree. Thus from the theory of client-centered therapy we have drawn hypotheses such as these: during therapy feelings which have previously been denied to awareness are experienced, and are assimilated into the concept of self; during therapy the concept of the self becomes more congruent with the concept of the ideal self; during and after therapy the observed behavior of the client becomes more socialized and mature; during and after therapy the client increases in attitudes of self-acceptance, and this is correlated with an increase in acceptance of others.

These are a few of the hypotheses we have been able to investigate. It will perhaps be clear that we have abandoned entirely the idea of one general criterion for our studies, and have substituted instead a number of clearly defined variables, each one specific to the hypothesis being investigated. This means that it was our hope in the research to be able to state our conclusions in some such form as this: that client-centered psychotherapy produces measurable changes in characteristics *a*, *b*, *d*, and *f*, for example, but does not produce changes in variables *c* and *e*. When statements of this sort are available then the professional worker and the layman will be in a position to make a value judgment as to whether he regards as a "success" a process which produces these changes. Such value judgments will not, however, alter the solid facts in our slowly growing scientific knowledge of the effective dynamics of personality change.

Thus in our research we have, in place of the usual global criterion of "success," many specific criterion variables, each drawn from our theory of therapy, and each operationally defined.

This resolution of the problem of criteria was of great help in making an intelligent selection of research instruments to use in our battery of test. We did not ask the unanswerable question as to what instruments would measure success or cure. We asked instead, specific questions related to each hypothesis. What instrument can be used to measure the individual's concept of self? What instrument will give a satisfactory measure of maturity of behavior? How can we measure the degree of an individual's acceptance of others? While questions such as these are difficult, operational answers are discoverable. Thus our decision in regard to criteria gave us much help in solving the whole problem of instrumentation of the research.

The Design of the Research

The fact that there has been no objective evidence of constructive personality change brought about by psychotherapy, has been mentioned by a number of thoughtful writers. Hebb states that "there is no body of fact to show that psychotherapy is valuable" (4, p. 271). Eysenck, after surveying some of the available studies, points out that the data "fail to prove that psychotherapy, Freudian or otherwise, facilitates the recovery of neurotic patients" (3, p. 322).

Mindful of this regrettable situation we were eager to set up our investigation in a sufficiently rigorous fashion that the confirmation or disproof of our hypotheses would establish two points: (a) that significant change had or had not occurred, and (b) that such change, if it did occur, was attributable to the therapy and not to some other factor. In such a complex field as therapy it is not easy to devise a research design which will accomplish these aims, but we believe that we have made real progress in this direction.

Having chosen the hypotheses which we wished to test, and the instruments most suitable for their operational measurement, we were now ready for the next step. This selected series of objective research instruments were used to measure various characteristics of a group of clients before their therapy, after the completion of therapy, and at a followup point six months to one year later, as indicated in Figure 1. The clients were roughly typical of those coming to the Counseling Center of the University of Chicago, and the aim was to collect this data, including the recording of all interviews, for at least 25 clients. The choice was made to make an intensive study of a group of moderate size, rather than a more superficial analysis of a larger number.

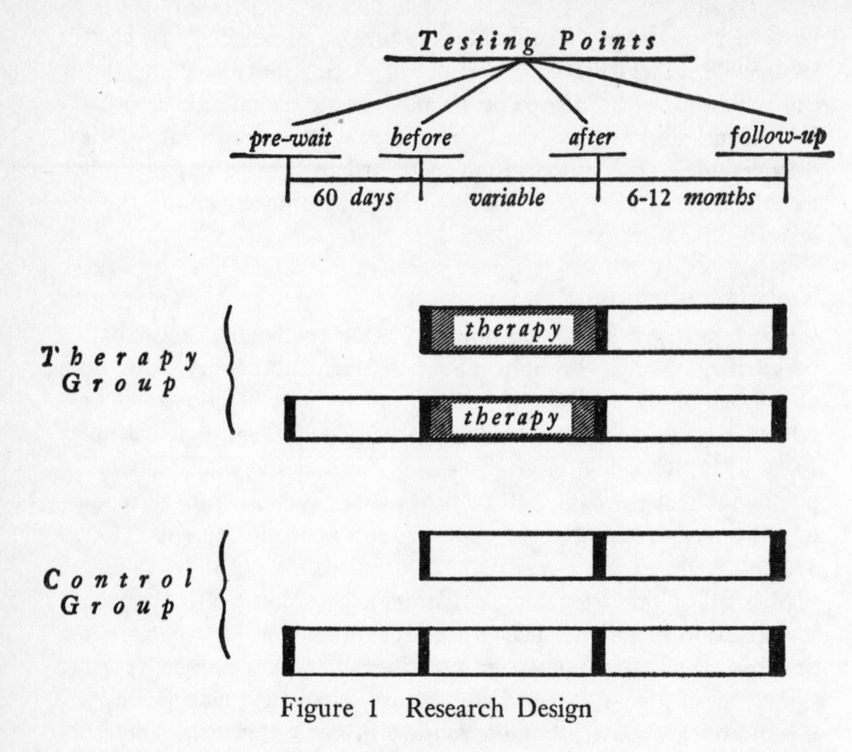

Figure 1 Research Design

A part of the therapy group was set aside as an own-control group. This group was given the battery of research instruments, asked to wait during a two month control period, and then given the battery a second time before counseling. The rationale of this procedure is that if change occurs in individuals simply because they are motivated for therapy, or because they have a certain type of personality structure, then such change should occur during this control *period.*

Another group of individuals not in therapy was selected as an equivalent-control group. This group was equivalent in age and age distribution to the therapy group, and roughly equivalent in socio-economic status, in the proportion of men and women, and of students and non-students. This group was given the same tests as the therapy group, at matched time intervals. A portion of this group was given the test battery four times, in order to make them strictly comparable to the own-control therapy group. The rationale of this equivalent-control group is that if change occurs in individuals as the result of the passage of time, or the influence of random variables, or as an artifact of the repeated administration of tests, then such change should be evident in the findings from this group.

The over-all logic of this doubly controlled design is that if the therapy group shows changes during and after the therapy period which are significantly greater than those which occur in the own-control *period* or in the equivalent-control *group*, then it is reasonable to attribute these changes to the influence of the therapy.

I cannot, in this brief report, go into the complex and ramified details of the various projects which were carried out within the framework of this research design. A more complete account (7) has been prepared which describes thirteen of the projects completed thus far. Suffice to say that complete data on 29 clients, dealt with by 16 therapists, was obtained, as well as complete data on a matched control group. The careful evaluation of the research findings enables us to draw certain conclusions such as these: That profound changes occur in the perceived self of the client during and after therapy; that there is constructive change in the client's personality characteristics and personality structure, changes which bring him closer to the personality characteristics of the well-functioning person; that there is change in directions defined as personal integration and adjustment; that there are changes in the maturity of the client's behavior as observed by friends. In each instance the change is significantly greater than that found in the control group or in the clients during the own-control period. Only in regard to the hypotheses having to do with acceptant and democratic attitudes in relation to others are the findings somewhat confused and ambiguous.

In our judgment, the research program which has already been completed has been sufficient to modify such statements as those made by Hebb and Eysenck. In regard to client-sponsored psychotherapy, at least, there is now objective evidence of positive changes in personality and behavior in directions which are usually regarded as constructive and these changes are attributable to the therapy. It is the adoption of multiple specific research criteria and the use of a rigorously controlled research design which makes it possible to make such a statement.

THE MEASUREMENT OF CHANGES IN THE SELF

Since I can only present a very small sample of the results, I will select this sample from the area in which we feel there has been the most significant advance in methodology, and the most provocative findings, namely, our attempts to measure the changes in the client's perception of himself, and the relationship of self-perception to certain other variables.

In order to obtain an objective indication of the client's self-perception, we made use of the newly devised Q-technique, developed by Stephenson (9). A large "universe" of self-descriptive statements was drawn from recorded interviews and other sources. Some typical statements are: "I am a submissive person"; "I don't trust my emotions"; "I feel relaxed and nothing bothers me"; "I am afraid of sex"; "I usually like people"; "I have an attractive personality"; "I am afraid of what other people think of me." A random sample of one hundred of these, edited for clarity, was used as the instrument. Theoretically we now had a sampling of all the ways in which an individual could perceive himself. These hundred statements, each printed on a card, were given to the client. He was asked to sort the cards to represent himself "as of now," sorting the cards into nine piles from those items most characteristic of himself to those least characteristic. He was told to place a certain number of items in each pile so as to give an approximately normal distribution of the items. The client sorted the cards in this way at each of the major points, before therapy, after, and at the followup point, and also on several occasions during therapy. Each time that he sorted the cards to picture himself he was also asked to sort them to represent the self he would like to be, his ideal self.

We thus had detailed and objective representations of the client's self-perception at various points, and his perception of his ideal self. These various sortings were then inter-correlated, a high correlation between two sortings indicating similarity or lack of change, a low correlation indicating a dissimilarity, or a marked degree of change.

In order to illustrate the way in which this instrument was used to test some of our hypotheses in regard to the self, I am going to present some of the findings from the study of one client (from 7, ch. 15) as they relate to several hypotheses. I believe this will indicate the provocative nature of the results more adequately than presenting the general conclusions from our study of self-perception, though I will try to mention these generalized results in passing.

The client from whose data I will draw material was a woman of 40, most unhappy in her marriage. Her adolescent daughter had had a nervous breakdown, about which she felt guilty. She was a rather deeply troubled person who was rated on diagnostic measures as seriously neurotic. She was not a member of the own-control group, so entered therapy immediately after taking the first battery of tests. She came for 40 interviews over a period of 5½ months, when she concluded therapy. Followup tests were administered seven months later, and at that time she decided to come in for 8 more interviews. A second followup study was done 5 months later. The counselor judged that there had been very considerable movement in therapy.

Figure 2 presents some of the data regarding the changing self-perception of this client. Each circle represents a sorting for the ideal self or the self. Sortings were done before therapy, after the seventh and twenty-fifth interviews, at the end of therapy, and at the first and second followup points. The correlations are given between many of these sortings.

Let us now examine this data in reference to one of the hypotheses which we were interested in testing, namely, that the perceived self of the client will change more during therapy than during a period of no therapy. In this particular case the change was greater during therapy ($r = .39$) than during either of the followup periods ($r = 74$, $.70$) or the whole twelve month period of followup ($r = .65$). Thus the hypothesis is upheld in this one case. In this respect she was characteristic of our clients, the general finding being that the change in the perceived self during therapy was significantly greater than during the control or followup periods, and significantly greater than the change occurring in the control group.

Let us consider a second hypothesis. It was predicted that during and after therapy the perceived self would be more positively valued,

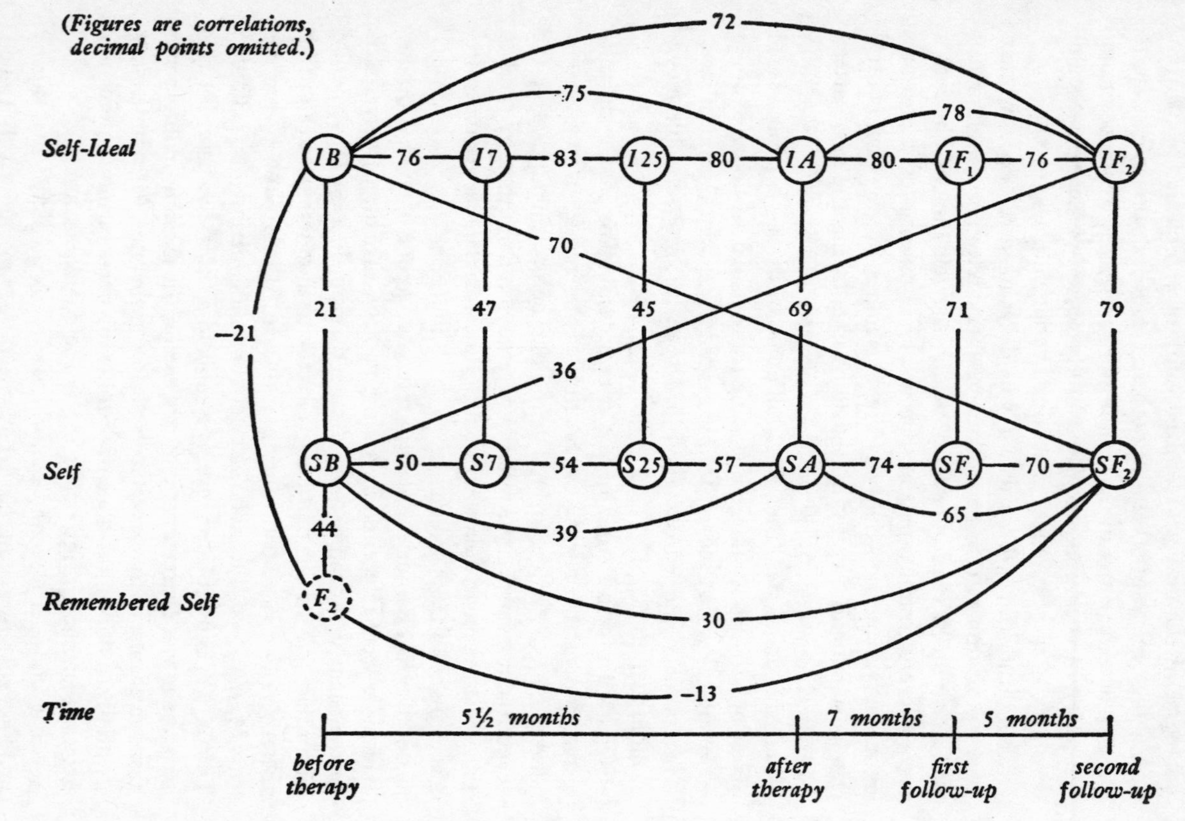

Figure 2 The changing relationships between Self and Self-Ideal

i.e., would become more congruent with the ideal, or valued, self.

This client exhibits considerable discrepancy between the self she is and the self she would like to be, when she first comes in ($r = .21$). During and after therapy this discrepancy decreases, a decided degree of congruence existing at the final followup study ($r = .79$), thus confirming our hypothesis. This is typical of our general findings, which showed a significant increase in congruence between self and ideal, during therapy, for the group as a whole.

Close study of Figure 2 will show that by the end of our study, the client perceives herself as having become very similar to the person she wanted to be when she came in ($rIB \cdot SF_2 = .70$). It may also be noted that her final self-ideal became slightly more similar to her initial self ($rSB \cdot IF_2 = .36$) than was her initial ideal.

Let us briefly consider another hypothesis, that the change in the perceived self will not be random, but will be in a direction which expert judges would term adjustment.

As one part of our study the Q-sort cards were given to a group of clinical psychologists not associated with the research, and they were asked to sort the cards as they would be sorted by a "well-adjusted" person. This gave us a criterion sorting with which the self-perception of any client could be compared. A simple score was developed to express the degree of similarity between the client's self-perception and this representation of the "adjusted" person. This was called the "adjustment score," higher scores indicating a higher degree of "adjustment."

In the case of the client we have been considering the adjustment scores for the six successive self-sorts shown in Figure 2, beginning with the self as perceived before therapy, and ending at the second followup point, are as follows: 35, 44, 41, 52, 54, 51. The trend toward improved adjustment, as operationally defined, is evident. This is also true for the group as a whole, a marked increase in adjustment score occurring over the period of therapy, and a very slight regression in score during the followup period. There was essentially no change in the control individuals. Thus, both for this particular client, and for the group as a whole, our hypothesis is upheld.

When a qualitative analysis of the different self-sorts is made, the findings further confirm this hypothesis. When the initial self-picture is compared with those after therapy, it is found that after therapy the client sees herself as changed in a number of ways. She feels she is more self-confident and self-reliant, understands herself better, has more inner comfort, and more comfortable relationships with others. She feels less guilty, less resentful, less driven and insecure, and feels less need for self-concealment. These qualitative changes are similar to those shown by the other clients in the research and are in general in accord with the theory of client-centered therapy.

I should like to point out certain additional findings of interest which are illustrated in Figure 2.

It will be evident that the representation of the ideal self is much more stable than the representation of the self. The inter-correlations are all above .70, and the conception of the person she would like to be changes relatively little over the whole period. This is characteristic of almost all of our clients. While we had formulated no hypothesis on this point it had been our expectation that some clients would achieve greater congruence of self and ideal primarily through alteration of their values, others through the alteration of self. Our evidence thus far indicates that this is incorrect, and that with only occasional exceptions, it appears to be the concept of the self which exhibits the greater change.

Some change, however, does occur in the ideal self in the case of our client and the direction of this slight change is of interest. If we calculate the previously described "adjustment score" of the successive representations of the ideal self of this client, we find that the average score for the first three is 57, but the average of the three following therapy is 51. In other words the self-ideal has become less perfectly "adjusted," or more attainable. It is to some degree a less punishing goal. In this respect also, this client is characteristic of the trend in the whole group.

Another finding has to do with the "remembered self" which is shown in Figure 2. This sorting was obtained by asking the client, at the time of the second followup study, to sort the cards once more to represent herself as she was when she first entered therapy. This remembered self turned out to be very different from the self-picture she had given at the time of entering therapy. It correlated only .44 with the self-representation given at that time. Furthermore, it was a much less favorable picture of her self, being far more discrepant from her ideal ($r = -.21$), and having a low adjustment score — a score of 26 compared to a score of 35 for the initial self-picture. This suggests that in this sorting for the remembered self we have a crude objective measure of the reduction in defensiveness which has occurred over the eighteen-month period of our study. At the final contact she is able to give a considerably truer picture of the maladjusted and disturbed person that she was when she entered therapy, a picture which is confirmed by other evidence, as we shall see. Thus the degree of alteration in the self over the total period of a year and a half is perhaps better represented by the correlation of −.13 between the remembered self and the final self, than by the correlation of .30 between the initial and final self.

Let us now turn to a consideration of one more hypothesis. In client-centered therapy our theory is that in the psychological safety of the therapeutic relationship the client is able to permit in his awareness feelings and experiences which ordinarily would be repressed, or denied to awareness. These previously denied experiences now become incorporated into the self. For example, a client who has repressed all feelings of hostility may come, during therapy, to experience his hostility freely. His concept of himself then becomes reorganized to include this realization that he has, at times, hostile feelings toward others. His self-picture becomes to that degree a more accurate map or representation of the totality of his experience.

We endeavored to translate this portion of our theory into an operational hypothesis, which we expressed in this way: During and after therapy there will be an increasing congruence between the self as perceived by the client and the client as perceived by a diagnostician. The assumption is that a skilled person making a psychological diagnosis of the client is more aware of the totality of the client's experience patterns, both conscious and unconscious, than is the client. Hence if the client assimilates into his own conscious self-picture many of the feelings and experiences which previously he has repressed, then his picture of himself should become more similar to the picture which the diagnostician has of him.

The method of investigating this hypothesis was to take the projective test (the Thematic Apperception Test) which had been administered to the client at each point and have these four tests examined by a diagnostician. In order to avoid any bias, this psychologist was not told the order in which the tests had been administered. He was then asked to sort the Q-cards for each one of the tests to represent the client as she diagnostically was at that time. This procedure gave us an unbiased diagnostic evaluation, expressed in terms of the same instrument as the client had used to portray herself, so that a direct and objective comparison was possible, through correlation of the different Q-sorts.

The result of this study, for this particular client, is shown in Figure 3. The upper portion of this diagram is simply a condensation of the information from Figure 2. The lowest row shows the sortings made by the diagnostician, and the correlations enable us to test our hypothesis. It will be observed that at the beginning of therapy there is no relationship between the client's perception of herself and the diagnostician's perception of the client (r = .00). Even at the end of therapy the situation is the same (r = .05). But by the time of the first followup (not shown) and the second followup, the client's perception of herself has become substantially like the diagnostician's perception of her (first followup, r = .56; second followup, r = .55). Thus the hypothesis is clearly upheld, congruence between the self as perceived by the client and the client as perceived by a diagnostician having significantly increased.

There are other findings from this aspect of the study which are of interest. It will be noted that at the time of beginning therapy the client as perceived by the diagnostician is very dissimilar to the ideal she had for herself (r = −.42). By the end of the study the diagnostician sees her as being decidedly similar to her ideal at that time (r = .46) and even more similar to the ideal she held for herself at the time she came in (r = .61). Thus we may say that the objective evidence indicates that the client has become, in her self-perception and in her total personality picture, substantially the person she wished to become when she entered therapy.

Another noteworthy point is that the change in the diagnostician's perception of the client is considerably sharper than is the change in the perceived self of the client (r = −.33, compared with r of

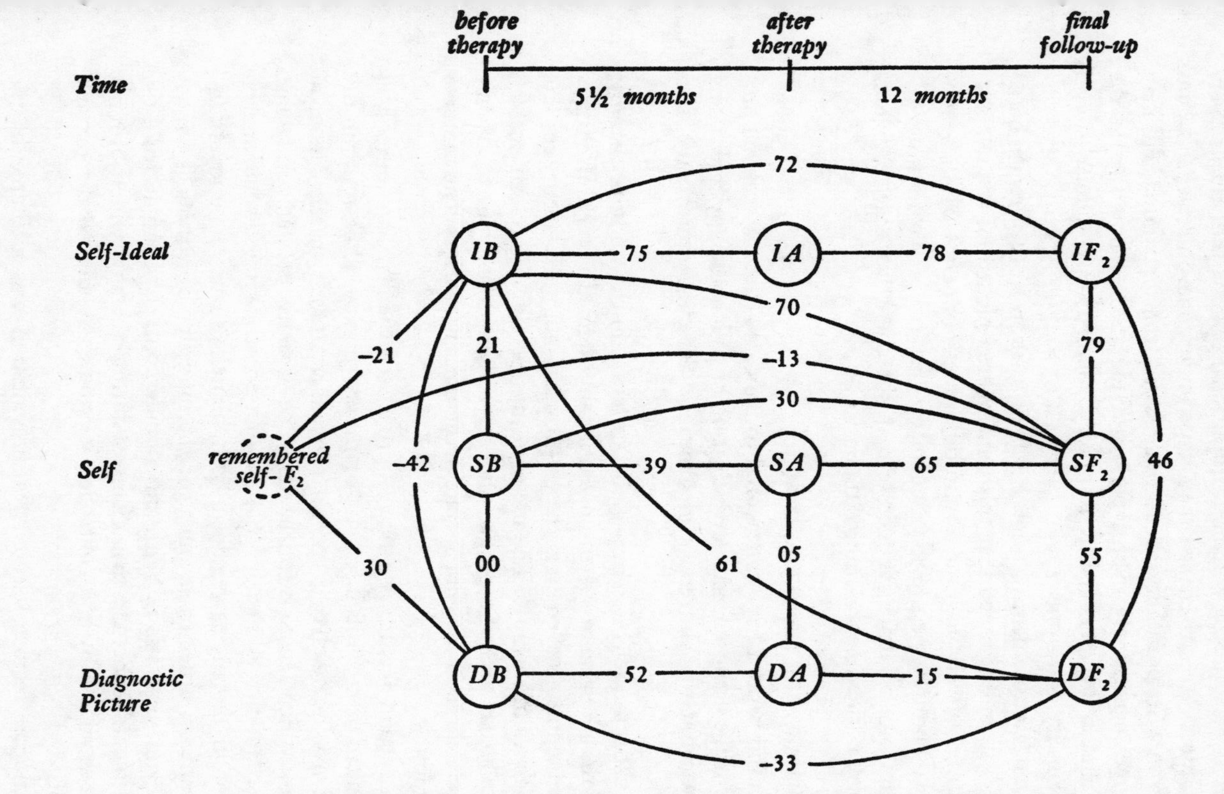

Figure 3 Relationship between Self, Self-Ideal, and Diagnosis
(Figures are correlations, decimal points omitted)

.30). In view of the common professional opinion that clients overrate the degree of change they have undergone, this fact is of interest. The possibility is also suggested that an individual may change so markedly over a period of eighteen months that at the conclusion his personality is more dissimilar than similar to his personality at the outset.

One last comment on Figure 3 is in relation to the "remembered self." It will be noted that this remembered picture of the self correlates positively with the diagnostic impression ($r = .30$), thus tending to confirm the previous statement that it represents a more accurate and less defensive picture than the client was able to give of herself at the time she entered therapy.

Summary and Conclusion

In this paper I have endeavored to indicate at least a skeleton outline of the comprehensive investigation of psychotherapy now going forward at the University of Chicago. Several features have been mentioned.

First is the rejection of a global criterion in the study of therapy, and the adoption of specific operationally defined criteria of change, based upon detailed hypotheses growing out of a theory of the dynamics of therapy. The use of many specific criteria has enabled us to make scientific progress in determining the types of change which do and do not occur concomitant with client-centered therapy.

A second feature is a new approach to the hitherto unresolved problem of controls in studies of psychotherapy. The research design has included two control procedures, (1) a matched control group which accounts for the influence of time, repeated test-taking, and random variables, and (2) an own-control group in which each client in therapy is matched with himself during a period of no therapy, in order to account for the influence of personality variables and motivation. With this double-control design it has been possible to conclude that changes during therapy which are not accounted for by the controlled variables, are due to the therapy itself.

Another feature selected for presentation was a sample of the progress which has been made in carrying on rigorous objective investigation of subtle elements of the client's subjective world. Evidence has been presented as to: the change in the self-concept of the client; the degree to which the perceived self becomes similar to the valued self; the extent to which the self as perceived becomes more comfortable and adjusted; and the degree to which the client's perception of self becomes more congruent with a diagnostician's perception of the client. These findings tend to confirm the theoretical formulations which have been made as to the place of the self-concept in the dynamic process of psychotherapy.

There are two conclusions which I would like to leave with you in closing. The first is that the research program I have described appears to make it quite clear that objective evidence, meeting the usual canons of rigorous scientific investigation, can be obtained as to the personality and behavioral changes brought about by psychotherapy, and has been obtained for one psychotherapeutic orientation. This means that in the future similar solid evidence can be obtained as to whether personality change occurs as a result of other psychotherapies.

The second conclusion is in my judgment even more significant. The methodological progress made in recent years means that the many subtleties of the therapeutic process are now wide open for research investigation. I have endeavored to illustrate this from the investigation of changes in the self-concept. But similar methods make it equally possible to study objectively the changing relationship between client and therapist, "transference" and "countertransference" attitudes, the changing source of the client's value system, and the like. I believe it may be said that almost any theoretical construct which is thought to be related to personality change or to the process of psychotherapy, is now amenable to research investigation. This opens a new vista of scientific investigation. The pursuit of this new path should throw much light on the dynamics of personality, particularly on the process of personality change in an interpersonal relationship.

REFERENCES

1. Axline, V. M. *Play Therapy*. Boston: Houghton Mifflin Co., 1947.
2. Curran, C. A. *Personality Factors in Counseling*. New York: Grune & Stratton, 1945.
3. Eysenck, H. J. The effects of psychotherapy: an evaluation, *J. Consult. Psychol.*, 1952, *16*, 319–324.
4. Hebb, D. O. *Organization of Behavior*. New York: Wiley, 1949.
5. Rogers, C. R. *Client-Centered Therapy*. Boston: Houghton Mifflin Co., 1951.
6. Rogers, C. R. *Counseling and Psychotherapy*. Boston: Houghton Mifflin Co., 1942.
7. Rogers, C. R., and R. Dymond, (Eds.). *Psychotherapy and Personality Change*. University of Chicago Press, 1954.
8. Snyder, W. U., (Ed.). *Casebook of Nondirective Counseling*. Boston: Houghton Mifflin Co., 1947.
9. Stephenson, W. U. *The Study of Behavior*. University of Chicago Press, 1953.

12

Client-Centered Therapy in Its Context of Research *

How could I make clear, to a European audience relatively unaccustomed to the American tradition of empirical research in psychology, the methods, the findings, the significance, of research in client-centered therapy? This was the task which was set for me by the fact that Dr. G. Marian Kinget and I were writing a book on client-centered therapy to be published first in Flemish and then in French. Dr. Kinget presented the clinical principles of such therapy. I presented the central theories of client-centered therapy (almost identical with the English presentation, A Theory of Therapy, Personality and Interpersonal Relationships, in S. Koch (ed.) Psychology: A Study of a Science, vol. III. *(New York: McGraw-Hill, 1959), 184–256). I now wished to introduce them to the research in which we had engaged to confirm or disconfirm our theories. This chapter (slightly modified for this volume) is the result, and I hope it may have meaning for Americans as well as Europeans.*

In one small matter I beg the reader's indulgence. Three para-

* This is the English version of Chapter XII of the volume *Psychotherapie en menselijke verhoudingen: Theorie en praktijk van de non-directieve therapie* by Carl R. Rogers & G. Marian Kinget, Utrecht, The Netherlands: (Uitgeverij Het Spectrum, 1960).

graphs describing the development and use of the Q-sort by which self-perception is measured, are almost identical with similar material in Chapter II. I left them in so that either chapter might be read independently without reference to the other.

This chapter goes back to the earliest of our research efforts, around 1940, and concludes with a description of several of the unfinished projects which are still challenging our best efforts in 1961. Thus I have tried to present at least a small sampling of more than a score of years of research effort.

THE STIMULATION OF RESEARCH

One of the most important characteristics of the client-centered orientation to therapy is that from the first it has not only stimulated research but has existed in a context of research thinking. The number and variety of the completed studies is impressive. In 1953 Seeman and Raskin described or mentioned nearly fifty research investigations having to do with client-centered therapy with adults, in their critical analysis of the trends and directions of such research (9). In 1957 Cartwright published an annotated bibliography of research and theory construction in client-centered therapy, and found it necessary to include 122 references (4). He, like Seeman and Raskin, omitted all references having to do with research in play therapy and group therapy of a client-centered nature. There seems then no question but that the theory and practice of client-centered therapy have set in motion a surprising number of objective empirical investigations. It seems reasonable to ask ourselves why.

In the first place the theory of client-centered therapy has been seen from the first not as dogma or as truth but as a statement of hypotheses, as a tool for advancing our knowledge. It has been felt that a theory, or any segment of a theory, is useful only if it can be put to test. There has been a sense of commitment to the objective testing of each significant aspect of our hypotheses, believing that the only way in which knowledge can be separated from individual prejudice and wishful thinking is through objective investigation.

To be objective such investigation must be of the sort that another investigator collecting the data in the same way and performing the same operations upon it, will discover the same or similar findings, and come to the same conclusions. In short we have believed from the first that the field of psychotherapy will be advanced by the open, objective testing of all hypotheses in ways which are publicly communicable and replicable.

A second reason for the stimulating effect of the client-centered approach upon research is the orienting attitude that scientific study can begin anywhere, at any level of crudity or refinement; that it is a direction, not a fixed degree of instrumentation. From this point of view, a recorded interview is a small beginning in scientific endeavor, because it involves greater objectification than the memory of an interview; a crude conceptualization of therapy, and crude instruments for measuring these concepts, are more scientific than no such attempt. Thus individual research workers have felt that they could begin to move in a scientific direction in the areas of greatest interest to them. Out of this attitude has come a series of instruments of increasing refinement for analyzing interview protocols, and significant beginnings have been made in measuring such seemingly intangible constructs as the self-concept, and the psychological climate of a therapeutic relationship.

This leads me to what I believe to be the third major reason for the degree of success the theory has had in encouraging research. The constructs of the theory have, for the most part, been kept to those which can be given operational definition. This has seemed to meet a very pressing need for psychologists and others who have wished to advance knowledge in the field of personality, but who have been handicapped by theoretical constructs which cannot be defined operationally. Take for example the general phenomena encompassed in such terms as the self, the ego, the person. If a construct is developed — as has been done by some theorizers — which includes those inner events not in the awareness of the individual as well as those in awareness, then there is no satisfactory way at the present time to give such a construct an operational definition. But by limiting the self-concept to events in awareness, the construct can be given increasingly refined operational definition through the

Q-technique, the analysis of interview protocols, etc., and thus a whole area of investigation is thrown open. In time the resulting studies may make it possible to give operational definition to the cluster of events not in awareness.

The use of operationally definable constructs has had one other effect. It has made completely unnecessary the use of "success" and "failure" — two terms which have no scientific usefulness — as criteria in studies of therapy. Instead of thinking in these global and ill-defined terms research workers can make *specific* predictions in terms of operationally definable constructs, and these predictions can be confirmed or disconfirmed, quite apart from any value judgment as to whether the change represents "success" or "failure." Thus one of the major barriers to scientific advance in this area has been removed.

Another reason for whatever effectiveness the system has had in mediating research, is that the constructs have generality. Because psychotherapy is such a microcosm of significant interpersonal relationship, significant learning, and significant change in perception and in personality, the constructs developed to order the field have a high degree of pervasiveness. Such constructs as the self-concept, or the need for positive regard, or the conditions of personality change, all have application to a wide variety of human activities. Hence such constructs may be used to study areas as widely variant as industrial or military leadership, personality change in psychotic individuals, the psychological climate of a family or a classroom, or the inter-relation of psychological and physiological change.

One final fortunate circumstance deserves mention. Unlike psychoanalysis, for example, client-centered therapy has always existed in the context of a university setting. This means a continual process of sifting and winnowing of the truth from the chaff, in a situation of fundamental personal security. It means being exposed to the friendly criticism of colleagues, in the same way that new views in chemistry or biology or genetics are subjected to critical scrutiny. Most of all it means that the theory and the technique are thrown open to the eager searching of younger minds. Graduate students question and probe; they suggest alternative formulations; they undertake empirical studies to confirm or to disprove the various theoretical hypotheses. This has helped greatly to keep the client-centered orientation an open and self-critical, rather than a dogmatic, point of view.

It is for reasons of this sort that client-centered therapy has built into itself from the first the process of change through research. From a limited viewpoint largely centered on technique, with no empirical verification, it has grown to a ramifying theory of personality and interpersonal relations as well as of therapy, and it has collected around itself a considerable body of replicable empirical knowledge.

The Early Period of Research

Objective investigations of psychotherapy do not have a long history. Up to 1940 there had been a few attempts to record therapeutic interviews electronically, but no research use had been made of such material. There had been no serious attempts to utilize the methods of science to measure the changes which were thought to occur in therapy. So we are speaking of a field which is still, relatively speaking, in its swaddling clothes. But a beginning has been made.

Sometime in 1940 a group of us at Ohio State University successfully recorded a complete therapeutic interview. Our satisfaction was great, but it quickly faded. As we listened to this material, so formless, so complex, we almost despaired of fulfilling our purpose of using it as the data for research investigations. It seemed almost impossible to reduce it to elements which could be handled objectively.

Yet progress was made. Enthusiasm and skill on the part of graduate students made up for the lack of funds and suitable equipment. The raw data of therapy was transformed by ingenious and creative thinking into crude categories of therapist techniques and equally crude categories of client responses. Porter analyzed the therapist's behavior in significant ways. Snyder analyzed client responses in several cases, discovering some of the trends which existed. Others were equally creative, and little by little the possibility of research in this field became a reality.

These early studies were often unsophisticated, often faulty in research design, often based upon inadequate numbers, but their contribution as an opening wedge was nonetheless great.

SOME ILLUSTRATIVE STUDIES

In order to give some feeling for the steadily growing stream of research several studies will be described in sufficient detail to give some notion of their methodology and their specific findings. The studies reported are not chosen because they are especially outstanding. They are representative of different trends in the research as it developed. They will be reported in chronological order.

THE LOCUS OF EVALUATION

In 1949 Raskin (5) completed a study concerned with the perceived source of values, or the locus of the evaluating process. This started from the simple formulation that the task of the counselor was not to think *for* the client, or *about* the client, but *with* the client. In the first two the locus of evaluation clearly resides in the counselor, but in the last the counselor is endeavoring to think and empathize with the client within the latter's own frame of reference, respecting the client's own valuing process.

The question Raskin raised was whether the client's perceived locus of evaluation changed during therapy. Putting it more specifically, is there a decrease in the degree to which his values and standards depend upon the judgments and expectations of others, and an increase in the extent to which his values and standards are based upon a reliance upon his own experience?

In order to study this objectively, Raskin undertook the following steps.

1. Three judges working independently were asked to select, in several recorded interviews, those statements which had to do with the source of the client's values and standards. It was found that there was more than 80 per cent agreement in the selection of such statements, indicating that the study was dealing with a discriminable construct.

2. Selecting 22 of these items to represent a wide range of source of values, Raskin gave these items to 20 judges, asking them to distribute these statements in four piles according to the continuum being studied, with equal-appearing intervals between the piles. Twelve of the items rated most consistently were used to form and illustrate a scale of locus of evaluation, with values from 1.0 to 4.0. Step 1 represented an unqualified reliance on the evaluations made by others. Step 2 included those instances in which there was a predominant concern with what others think, but some dissatisfaction with this state of dependence. Step 3 represented those expressions in which the individual showed as much respect for his own valuing process as for the values and expectations of others, and showed an awareness of the difference between self-evaluation and dependence on others' values. Step 4 was reserved for those instances in which there was clear evidence of reliance upon one's own experience and judgment as the basic source of values.

An example illustrating stage 3 may give a more vivid picture of this scale. The following client statement was rated as belonging in this step of the scale.

"So I've made a decision that I wonder if it is right. When you're in a family where your brother has gone to college and everybody has a good mind, I wonder if it is right to see that I am as I am and I can't achieve such things. I've always tried to be what the others thought I should be, but now I'm wondering whether I shouldn't just see that I am what I am." (6, p. 151).

3. Raskin now used this scale to rate each of 59 interviews in ten brief but fully recorded cases which had been made the subject of other research investigations. After he had made these ratings, but before analyzing them, he wished to determine the reliability of his judgments. Consequently he chose at random one item relating to locus of evaluation from each of the 59 interviews, and had these rated independently by another judge who knew nothing of the source of the items, or whether they came from early or late interviews. The correlation between the two sets of ratings was .91, a highly satisfactory reliability.

4. Having constructed a scale of equal-appearing intervals, and having demonstrated that it was a reliable instrument, Raskin was

now ready to determine whether there had been any shift in the locus of evaluation during therapy. The average score for the first interviews in the ten cases was 1.97, for the final interviews 2.73, a difference significant at the .01 level. Thus the theory of client-centered therapy on this point was upheld. A further confirmation was available. These 10 cases had been studied in other objective ways, so that there were objective criteria from other studies as to which cases were more, and which less successful. If one takes the five cases judged as more suiiessful, the shift in locus of evaluation in these cases is even sharper, the average for the first interviews being 2.12, and for the final interviews 3.34.

This study is, in a number of respects, typical of a large group of the research investigations which have been made. Starting with one of the hypotheses of client-centered theory, an instrument is devised to measure varying degrees of the construct in question. The instrument is then itself put to the test to determine whether it does in fact measure what it purports to measure, and whether any qualified person can use it and obtain the same or similar results. The instrument is then applied to the data of therapy in a way which can be shown to be unbiassed. (In Raskin's case the checking of 59 randomly selected items by another judge shows that bias, conscious or unconscious, did not enter appreciably into his ratings.) The data acquired from the use of the instrument can then be analyzed to determine whether it does or does not support the hypothesis. In this case the hypothesis was upheld, confirming the theory that clients in client-centered therapy tend to decrease in the extent to which they rely for guidance upon the values and expectations of others, and that they tend to increase in reliance upon self-evaluations based upon their own experiences.

Although the number of cases studied is small, and the therapy very brief (as was characteristic of that earlier period) these are the only major flaws in this study. It is probable that if replicated on a larger number of longer cases the results would still be the same. It marks an intermediate level of research sophistication, somewhere between the very crude initial studies, and the more meticulously designed recent studies.

The Relation of Autonomic Function to Therapy

Thetford undertook a study of quite a different sort, also completed in 1949 (11). His hypothesis went well beyond the theory of client-centered therapy, predicting physiological consequences which were consistent with the theory, but which had never been formulated.

Briefly his major hypothesis was that if therapy enables the individual to reorient his pattern of life and to reduce the tension and anxiety he feels regarding his personal problems, then the reactions of his automatic nervous system in, for example, a situation of stress, should also be altered. Essentially he was hypothesizing that if a change in life pattern and in internal tension occurred in therapy, this should show up in organismic changes in autonomic functioning, an area over which the individual has no conscious control. Essentially he was asking, How deep are the changes wrought by client-centered therapy? Are they deep enough to affect the total organismic functioning of the individual?

Although his procedure was decidedly complex, it can be described simply enough in its essentials. A therapy group of nineteen individuals was recruited, composed of clients coming to the Counseling Center of the University of Chicago for personal help. They were invited to volunteer for a research in personality. Since all who were invited participated, except a few who could not arrange testing appointments, this was a representative group of student clients from the Center. Ten individuals went into individual therapy, three into individual and group therapy concurrently, and six into group therapy. A control group of seventeen individuals not in therapy was recruited, roughly similar in age and educational status to the therapy group.

Every individual, whether therapy or control, went through the same experimental procedure. The most significant aspects were these. The individual was connected by suitable electrodes to a polygraph which recorded his palmar skin conductance (GSR), heart rate, and respiration. After a rest period to establish a base line, the individual was told that memory for digits was a good

index of intelligence, and that the experimenter wished to test him for this. The series of digits used was increased in length until the individual clearly failed. After a two minute rest, another series was used to bring another clear failure. After another rest, there was another frustrating failure. Since these were all students, the ego-involvement and the frustration were clearly real since the experience seemed to cast doubt on their intellectual ability. After another rest period the individual was released, but informed that he would be called back at a later time. At no time was there any hint that the experiment had anything to do with the individual's therapy, and the testing was carried on in another building.

Following the completion of therapy the clients were recalled and went through the same experimental procedure — three episodes of frustration and recovery, with continuous autonomic measurements being made. At matched time intervals, the controls were also recalled and put through an identical procedure.

Various physiological indices were computed for the therapy and control groups. The only significant differences between the groups were differences in the rapidity of recovery from frustration on the pre as compared with the post test. In general it may be said that the group which had therapy recovered from its frustration more quickly on the post-test than on the pre-test, while for the control group the results were the reverse. They recovered more slowly at the time of the second series of frustrations.

Let me make this more specific. The therapy group showed a change in the "recovery quotient" based on the GSR which was significant at the .02 level of confidence, and which was in the direction of more rapid recovery from frustration. The control group showed a change in the "recovery quotient" which was significant at the 10 per cent level, and was in the direction of a slower recovery. In other words they were less able to cope with the frustration during the post-test than during the pre-test. Another GSR measure, "per cent of recovery," again showed the therapy group making a more rapid recovery at the second test, a change significant at the 5 per cent level, while the control group showed no change. As to cardiovascular activity the therapy group, on the average, showed less heart-rate variation at the time of the

post-test frustration, a change significant at the 5 per cent level. The control group showed no change. Other indices showed changes consistent with those mentioned, but not as significant.

In general it may be said that the individuals who had experienced therapy developed a higher frustration threshold during their series of therapeutic contacts, and were able to recover their homeostatic balance more rapidly following frustration. In the control group, on the other hand, there was a slight tendency toward a lower threshold for the second frustration, and a definitely less rapid recovery of homeostasis.

In simple terms, the significance of this study appears to be that after therapy the individual is able to meet, with more tolerance and less disturbance, situations of emotional stress and frustration; that this description holds, even though the particular frustration or stress was never considered in therapy; that the more effective meeting of frustration is not a surface phenomenon but is evident in autonomic reactions which the individual cannot consciously control and of which he is completely unaware.

This study of Thetford's is characteristic of a number of the more pioneering and challenging of those which have been carried on. It went beyond client-centered theory as it had been formulated, and made a prediction consistent with the theory, and perhaps implicit in it, but well beyond the limits of the theory as it stood. Thus it predicted that if therapy enabled the individual better to handle stress at the psychological level, then this should be evident also in his autonomic functioning. The actual research was the testing of the correctness of the prediction. There is no doubt that the confirming effect on the theory is somewhat greater when rather remote predictions are tested and found to be correct.

Client Response to Differing Techniques

A small study completed by Bergman (2) in 1950 is an example of the way in which recorded interviews lend themselves to microscopic studies of the therapeutic process. He wished to study the question, What is the nature of the relationship between the counselor's method or technique and the client's response?

He chose to study all the instances in ten recorded cases (the same cases studied by Raskin and others) in which the client requested an evaluation from the counselor. There were 246 such instances in the ten cases, in which the client requested some solution for his problems, or an evaluation of his adjustment or progress, or a confirmation of his own view, or a suggestion as to how he should proceed. Each of these instances was included in the study as a response unit. The response unit consisted of the total client statement which included the request, the immediately following response by the counselor, and the total client expression which followed the counselor statement.

Bergman found that the counselor responses to these requests could be categorized in the following ways.

1. An evaluation-based response. This might be an interpretation of the client material, agreement or disagreement with the client, or the giving of suggestions or information.

2. A "structuring" response. The counselor might explain his own role, or the way in which therapy operates.

3. A request for clarification. The counselor might indicate that the meaning of the client's request is not clear to him.

4. A reflection of the context of the request. The counselor might respond by trying to understand the client material encompassing the request, but with no recognition of the request itself.

5. A reflection of the request. The counselor might endeavor to understand the client's request, or the client's request in a context of other feelings.

Bergman developed the following categories to contain the client expression subsequent to the counselor response.

1. Client again presents a request for evaluation, either a repetition of the same request or some enlargement or modification of it, or another request.

2. Client, whether accepting or rejecting the counselor response, abandons the attempt to explore his attitudes and problems (usually going off into other less relevant material.)

3. Client continues to explore his attitudes and problems.

4. Client verbalizes an understanding of relationships between feelings — expresses an insight.

Having checked the reliability of this categorization of both client and counselor material and having found it satisfactory, Bergman proceeded to analyze his data. He determined which categories occurred in conjunction with other categories more frequently than could be accounted for by chance. Some of the significant findings are these.

There was essentially only a chance relationship between the categories of initial client request and subsequent client response. The same was true of initial client request and counselor response. Thus neither the counselor's response nor the client's subsequent expression seemed to be "caused" by the initial request.

On the other hand there was significant interaction found between the counselor's response and the client's subsequent expression.

1. Reflection of feeling by the counselor is followed, more often than would be expected by chance, by continued self-exploration or insight. This relationship is significant at the 1 per cent level.
2. Counselor responses of types 1 and 2 (evaluation-based and interpretive responses or "structuring" responses) are followed, more often than would be expected by chance, by abandonment of self-exploration. This too is significant at the 1 per cent level.
3. A counselor response requesting clarification tends to be followed by repetition of the request, or by a decrease in self-exploration and insight. These consequences are significant at the 1 per cent and 5 per cent level, respectively.

Thus Bergman concludes that self-exploration and insight, positive aspects of the therapeutic process, appear to be furthered primarily by responses which are "reflections of feeling," while evaluative, interpretive, and "structuring" responses tend to foster client reactions which are negative for the process of therapy.

This study is an illustration of the way in which, in a number of investigations, the verbal recording of therapeutic interviews has been examined in a very minute and molecular way, in order to cast light upon some aspect of client-centered theory. In these studies the internal events of therapy have been examined objectively for the light they can throw upon the process.

A STUDY OF THE SELF-CONCEPT

Many investigations have been made of the changes in the client's concept of self, a construct which is central to the client-centered theory of therapy and personality. One, a study by Butler and Haigh (3), will be briefly reported here.

A method which has frequently been used for this purpose is the Q-technique developed by Stephenson (10), and adapted for the study of the self. Since an instrument based on this technique is used in the Butler and Haigh study, it may be simply described before giving the findings of the study itself.

From a number of recorded counseling cases a large population of all the self-referent statements was gleaned. From this a selection of 100 statements was made, and the statements edited for the sake of clarity. The aim was to select the widest possible range of ways in which the individual could perceive himself. The list included such items as: "I often feel resentful"; "I am sexually attractive"; "I really am disturbed"; "I feel uncomfortable while talking with someone"; "I feel relaxed and nothing really bothers me."

In the Butler and Haigh study each person was asked to sort the cards containing the 100 items. First he was to "Sort these cards to describe yourself as you see yourself today." He was asked to sort the cards into nine piles, from those most unlike him, to those most like him. He was asked to place a certain number in each pile. (The numbers in each pile were 1, 4, 11, 21, 26, 21, 11, 4, 1, thus giving a forced and approximately normal, distribution.) When he had completed this sort he was asked to sort the cards once more "to describe the person you would most like within yourself to be." This meant that for each item one would obtain the individual's self-perception, and also the value he attached to this characteristic.

It will be evident that the various sortings can be correlated. One can correlate the self pre-therapy with the self post-therapy, or the self with the ideal self, or the ideal self of one client with the ideal of another. High correlations indicate little discrepancy or change, low correlations the reverse. Study of the specific items which have been changed in their placement over therapy, for example, gives a qualitative picture of the nature of the change. Because of the large population of items there is less loss of clinical richness in the statistical investigation. By and large this procedure has enabled investigators to turn subtle phenomenological perceptions into objective and manipulable data.

Let us turn to the use made of the Q-sort of self items in the Butler and Haigh study. The hypotheses were: (1) that client-centered therapy results in a decrease in the discrepancy between the perceived self and the valued self; and (2) that this decrease in discrepancy will be more marked in clients who have been judged, on the basis of independent criteria, as having exhibited more movement in therapy.

As part of a much more comprehensive total program of research (8) the Q-sort for self and for ideal self was given to 25 clients before therapy started, after the conclusion of therapy, and at a follow-up point six to twelve months after the conclusion of therapy. The same program of testing was followed in a non-therapy control group matched for age, sex and socio-economic status.

The findings are of interest. The self-ideal correlations in the client group before therapy ranged from −.47, a very marked discrepancy between self and ideal, to .59, indicating that the self is quite highly valued as it is. The mean correlation at pre-therapy was −.01. At the conclusion of therapy the mean was .34, and at the follow-up point it was .31. This represents a highly significant change, supporting the hypothesis. It is of special interest that the correlation decreases only very slightly during the follow-up period. When attention is directed to the 17 cases who on the basis of counselor ratings and change on the Thematic Apperception Test had shown the most definite improvement in therapy, the change is even sharper. Here the mean at pre-therapy was .02, at follow-up time, .44.

Fifteen members of the group constituted an "own-control" group. They had been tested when they first requested help, then asked to wait for 60 days before beginning therapy. They were re-tested at the end of the 60-day period, as well as at the post-therapy and follow-up times. In this group of fifteen the self-ideal correlation at the first test was −.01 and at the end of the 60-day period it was identical, −.01. Thus the change which occurred during therapy is clearly associated *with* therapy, and does not result simply from the passage of time, or from a determination to obtain help.

The control group showed a very different picture from the therapy clients. The initial correlation of self and ideal was .58, and this did not change, being .59 at the follow-up point. Obviously this group did not feel the tension felt by the client group, tended to value themselves, and did not change appreciably in this respect.

It is reasonable to conclude from this study that one of the changes associated with client-centered therapy is that self-perception is altered in a direction which makes the self more highly valued. This change is not a transient one, but persists after therapy. This decrease in internal tension is a highly significant one, but even at the end of therapy the self is somewhat less valued than is found to be the case in a non-therapy control group. (Therapy, in other words, has not brought about "perfect adjustment," or a complete absence of tension.) It is also clear that the changes under discussion have not occurred simply as a result of the passage of time, nor as the result of a decision to seek help. They are definitely associated with the therapy.

This study is an example of many which have thrown light on the relationship of therapy to self-perception. From other studies (reported in Rogers and Dymond (8)) we know that it is primarily the self-concept which changes in therapy, not the ideal self. The latter tends to change but slightly, and its change is in the direction of becoming a less demanding, or more achievable self. We know that the self-picture emerging at the end of therapy is rated by clinicians (in a manner which excludes possible bias) as being better adjusted. We know that this emerging self has a greater degree of inner comfort, of self-understanding and self-acceptance, of self-responsibility. We know that this post-therapy self finds greater satisfaction and comfort in relationships with others. Thus bit by bit we have been able to add to our objective knowledge of the changes wrought by therapy in the client's perceived self.

DOES PSYCHOTHERAPY BRING CHANGE IN EVERYDAY BEHAVIOR?

The studies described thus far in this chapter, and others which might be cited, provide evidence that client-centered therapy brings many changes. The individual makes choices and establishes values differently; he meets frustration with less prolonged physiological tension, he changes in the way he perceives himself and values himself. But this still leaves unanswered the question of practical concern to the layman and to society, "Does the client's everyday behavior change in such a way that the changes can be observed, and is the nature of these changes positive?" It was to try to answer this question that I, with the help of colleagues, undertook an investigation of changes in the maturity of the client's behavior as related to therapy, a study published in 1954 (6).

The theory of client-centered therapy hypothesizes that the inner changes taking place in therapy will cause the individual after therapy to behave in ways which are less defensive, more socialized, more acceptant of reality in himself and in his social environment, and which give evidence of a more socialized system of values. He will, in short, behave in ways which are regarded as more mature, and infantile ways of behaving will tend to decrease. The difficult question to which we addressed ourselves was whether an operational definition could be given to such a hypothesis in order to put it to empirical test.

There are few instruments which even purport to measure the quality of one's everyday behavior. The best for our purposes was that developed by Willoughby a number of years ago, and termed the Emotional Maturity Scale. He constructed many items descriptive of behavior and had them rated by 100 clinical workers — psychologists and psychiatrists — as to the degree of maturity they represented. On the basis of these judgments he selected 60 items to compose his Scale. The scores range from 1 (most immature) to 9 (most mature). Several of the items, and their score values, are listed below to give the reader something of the flavor of the Scale.

Score *Item*

1. S (subject) characteristically appeals for help in the solution of his problems (Item 9).
3. When driving an automobile, S is unperturbed in ordinary situations but becomes angry with other drivers who impede his progress (Item 12).

5. On unmistakable demonstration of his inferiority in some respect, S is impressed but consoles himself by the contemplation of those activities in which he is superior (Item 45).
7. S organizes and orders his efforts in pursuing his objective, evidently regarding systematic method as a means of achieving them (Item 17).
9. S welcomes legitimate opportunities for sexual expression; is not ashamed, fearful, or preoccupied with the topic (Item 53).

Having selected our instrument we were able to state our hypothesis in operational form: Following the completion of client-centered therapy, the behavior of the client will be rated, by himself and by others who know him well, as being more mature, as evidenced by a higher score on the E–M Scale.

The method of the study was necessarily complex, since accurate and reliable measurements of everyday behavior are difficult to obtain. The study was made as a part of a larger program of investigation of nearly thirty clients and an equal group of matched controls (8). The various steps were as follows.

1. The client, prior to therapy, was asked to make a self-evaluation of his behavior on the E–M Scale.

2. The client was asked for the names of two friends who knew him well and who would be willing to make ratings of him. The contact with these friends was by mail, and their ratings on the E–M Scale were mailed directly to the Counseling Center.

3. Each friend was requested to rate, at the same time that he rated the client, one other person well known to him. The purpose of this was to determine the reliability of the friend's ratings.

4. That half of the therapy group which had been designated as the own-control group, filled out the E–M Scale when first requesting help and again, sixty days later, before therapy began. Ratings of the client by his two friends were also obtained at each of these times.

5. At the conclusion of therapy the client and his two friends were again requested for a rating on the E–M Scale.

6. Six to twelve months following the conclusion of therapy rat-

ings of his behavior were again obtained from the client and his friends.

7. The members of the matched control group rated their behavior on the E–M Scale at each of the points from which such ratings were obtained from the therapy group.

This design assembled a large body of data permitting analysis from various angles. Only the major findings will be reported here.

The E–M Scale proved to have satisfactory reliability when used by any one rater, whether the client or an observer-friend. However the agreement between the different raters was not close.

The individuals in the matched non-therapy control group showed no significant change in their behavior ratings during any of the periods involved in the study.

The clients who were members of the own-control group showed no significant behavioral change during the sixty-day waiting period, whether judged by their own ratings or that of their friends.

There was no significant change in the observer's ratings of the client's behavior over the period of therapy or the combined period of therapy and follow-up. This was, of course, contrary to our hypothesis. It seemed desirable to determine whether this negative finding held for all clients regardless of the movement they appeared to make in therapy. Consequently the clients were divided into those rated by counselors as showing most, moderate, or least movement in therapy.

It was found that for those rated as showing the most movement in therapy, the friend's ratings of the client's maturity of behavior increased significantly (5 per cent level). In the group showing moderate movement there was little change, and in the group showing least movement there was a negative change, in the direction of less mature behavior.

There was a definite and significant correlation between the therapist's ratings of movement in therapy, and the friends' observations of change in everyday behavior. This correlation is particularly interesting because the therapist's judgment was based solely on client reactions in the therapy hour, with little or no knowledge of outside behavior. The friends' ratings were based solely on outside observation, with no knowledge of what was going on in therapy.

In general these findings were paralleled by the clients' ratings of their own behavior, with one interesting exception. Those clients who were rated by their counselors as showing movement in therapy rated themselves as showing an increase in maturity, the ratings being almost identical with those made by the observers. But those clients who were rated by the counselors as being least successful in therapy, and who were rated by observers as showing a deterioration in the maturity of behavior, described themselves in ways that gave them a sharp increase in maturity score both at the post-therapy and follow-up points. This seems to be clear evidence of a defensive self-rating when therapy has not gone well.

In general then the conclusion appears justified that where client-centered therapy has been judged to show progress or movement, there is a significant observable change in the client's everyday behavior in the direction of greater maturity. Where the therapist feels that there has been little or no movement in therapy, then some deterioration in behavior is observed, in the direction of greater immaturity. This last finding is of particular interest because it is the first evidence that disintegrative consequences may accompany unsuccessful efforts to obtain help in a relationship with a client-centered therapist. While these negative consequences are not great, they nevertheless warrant further study.

This research illustrates the efforts made to investigate various behavioral results of psychotherapy. It also suggests some of the many difficulties involved in planning a sufficiently rigorous design such that one can be sure that (a) behavioral changes did in fact occur, and (b) that such changes are a consequence of the therapy and not of some other factor.

Having made this global study of everyday behavior changes, it seems possible that further research on this topic might better be carried on in the laboratory, where changes in problem-solving behavior, adaptive behavior, response to threat or frustration, etc., might be carried on under better-controlled conditions. The reported study is however a pioneering one in indicating both that successful therapy produces positive behavioral change, and that unsuccessful therapy can produce negative changes in behavior.

The Quality of the Therapeutic Relationship as Related to Movement in Therapy

The final study I wish to report is one recently completed by Barrett-Lennard (1). He started from the theoretical formulation of mine regarding the necessary conditions for therapeutic change. He hypothesized that if five attitudinal conditions were present in the relationship, therapeutic change would occur in the client. To investigate this problem he developed a Relationship Inventory which had different forms for client and therapist, and which was designed to study five dimensions of the relationship. Thus far he has analyzed only the data from the client perceptions of the relationship, and it is these findings which I shall report.

In a fresh series of cases, in which he knew that he would have various objective measures of degree of change, Barrett-Lennard gave his Relationship Inventory to each client after the fifth interview. In order to give more of the flavor of his study, I will give several of the items regarding each variable.

He was interested, for example, in measuring the extent to which the client felt himself to be empathically understood. So he included items such as these regarding the therapist, to be rated by the client on a six-point scale from very true to very strongly not true. It will be evident that these represent different degrees of empathic understanding.

> He appreciates what my experience feels like to *me*.
>
> He tries to see things thru my eyes.
>
> Sometimes he thinks that I feel a certain way because he feels that way.
>
> He understands what I say from a detached, objective point of view.
>
> He understands my words but not the way I feel.

A second element he wished to measure was the *level* of regard, the degree of liking of the client by the therapist. To measure this there were items like the following, each one again rated from strongly true, to strongly not true.

He cares about me.
He is interested in me.
He is curious about "what makes me tick," but not really interested in me as a person.
He is indifferent to me.
He disapproves of me.

To measure the unconditionality of the regard, the extent to which there were "no strings attached" to the counselor's liking, items of this sort were included.

Whether I am expressing "good" feelings or "bad" ones seems to make no difference to the way he feels toward me.
Sometimes he responds to me in a more positive and friendly way than he does at other times.
His interest in me depends on what I am talking to him about.

In order to measure the congruence or genuineness of the therapist in the relationship, items of this sort were used.

He behaves just the way that he *is*, in our relationship.
He pretends that he likes me or understands me more than he really does.
There are times when his outward response is quite different from his inner reaction to me.
He is playing a role with me.

Barrett-Lennard also wished to measure another variable which he regarded as important — the counselor's psychological availability, or willingness to be known. To measure this, items of this sort were used.

He will freely tell me his own thoughts and feelings, when I want to know them.
He is uncomfortable when I ask him something about himself.
He is unwilling to tell me how he feels about me.

Some of his findings are of interest. The more experienced of his therapists were perceived as having more of the first four qualities than the less experienced therapists. In "willingness to be known," however, the reverse was true.

In the more disturbed clients in his sample, the first four measures all correlated significantly with the degree of personality change as objectively measured, and with the degree of change as rated by the therapist. Empathic understanding was most significantly associated with change, but genuineness, level of regard, and unconditionality of regard were also associated with successful therapy. Willingness to be known was not significantly associated.

Thus we can say, with some assurance, that a relationship characterized by a high degree of congruence or genuineness in the therapist; by a sensitive and accurate empathy on the part of the therapist; by a high degree of regard, respect, liking for the client by the therapist; and by an absence of conditionality in this regard, will have a high probability of being an effective therapeutic relationship. These qualities appear to be primary change-producing influences on personality and behavior. It seems clear from this and other studies that these qualities can be measured or observed in small samples of the interaction, relatively early in the relationship, and yet can predict the outcome of that relationship.

This study is an example of recent work which puts to test ever more subtle aspects of the theory of client-centered therapy. It is to be noted that this study does not deal with matters of technique or conceptualizations. It cuts through to intangible attitudinal and experiential qualities. Research in psychotherapy has, in my judgment, come a long way to be able to investigate such intangibles. The positive evidence in regard to four of the variables, and the lack of positive evidence in regard to the fifth variable, is to me an indication that helpful and discriminative findings may come from studies carried on at this level.

It is of more than passing interest that the relationship qualities associated with progress in therapy are all attitudinal qualities. While it may be that degree of professional knowledge, or skills and techniques will also be found to be associated with change, this study raises the challenging possibility that certain attitudinal and experiential qualities by themselves, regardless of intellectual knowledge or medical or psychological training, may be sufficient to stimulate a positive therapeutic process.

This investigation is a pioneering one in still another respect. It is one of the first explicitly designed to study the *causative* or change-producing elements of psychotherapy. In this respect theory has advanced sufficiently, and methodological sophistication as well, that we may look forward to an increasing number of investigations into the dynamics of personality change. We may in time be able to distinguish and measure the conditions which cause and produce constructive change in personality and behavior.

SOME CURRENT RESEARCH

Investigations relating to psychotherapy are burgeoning in the United States. Even the psychoanalytic group is embarking on several objective studies of the process of analytic therapy. It would be quite impossible to review what is going on today, since the picture is so complex, and so rapidly changing. I shall limit myself to very brief sketches of several research projects and programs related to client-centered therapy of which I have personal knowledge.

A study is going on at the University of Chicago under the direction of Dr. John Shlien to investigate the changes which occur in brief time-limited therapy, and to compare these changes with those which occur in the usual unlimited therapy. Clients are offered a definite number of interviews (twenty in most instances, forty in some) and therapy is concluded at the end of this time. Both the way in which individuals are able to use time, and the possibility of shortening the therapy period, are of interest to the investigators. This program should be completed in the not-too-distant future.

A study which is closely related is an investigation of short-term Adlerian therapy. With the active cooperation of Dr. Rudolph Dreikurs and his colleagues, Dr. Shlien is carrying on a study of Adlerian therapy exactly parallel to the above. If all goes well with the program it will mean that a direct comparison can be made of two sharply divergent therapies — Adlerian and client-centered — in which the same pre-tests and post-tests will have been administered, the therapy will be identical in length, and all interviews will have been recorded. This will indeed be a milestone, and should greatly expand our knowledge of the common and divergent elements in different forms of therapy.

Another study at the University of Chicago is being carried on by Dr. Desmond Cartwright, Donald Fiske, William Kirtner, and others. It is attempting to investigate, on a very broad basis indeed, a great many of the factors which may be associated with therapeutic change. It is casting a broad net to investigate many elements not previously considered which may be related to progress or lack of progress in therapy.

At the University of Wisconsin, Dr. Robert Roessler, Dr. Norman Greenfield, Dr. Jerome Berlin and I have embarked upon a ramified group of studies which it is hoped will, among other things, throw light on the autonomic and physiological correlates of client-centered therapy. In one portion of the investigation continuous recordings of GSR, skin temperature, and heart rate are being made on clients during the therapy hour. The comparison of these with the recorded interviews will perhaps give more information as to the fundamental physiological-psychological nature of the process of personality change.

A smaller project in which several individuals are at work involves the objective study of the process of psychotherapy. In a recent paper (7) I formulated a theoretical picture, based upon observation, of the irregularly sequential stages in the process of psychotherapy. We are currently at work translating this theoretical description into an operational scale which may be used to study recorded therapeutic interviews. Currently studies having to do with the reliability and validity of this scale are being carried on.

Still another program at the University of Wisconsin in which Dr. Eugene Gendlin and I are the principal investigators, concerns itself with a comparison of the process of psychotherapy in schizophrenic patients (both chronic and acute) with that in normal individuals. Each therapist in the study will take on three clients at a time, matched for age, sex, socio-educational status — one chronic schizophrenic, one acute schizophrenic, and one person of "normal" adjustment from the community. With a variety of pre-tests and post-tests, and a recording of all interviews, it is hoped that this study will have many findings of interest. It pushes the testing of client-centered hypotheses into a new field, that of the hospitalized psychotic person. Part of the fundamental hypothesis of the study is that given the necessary conditions of therapy (somewhat as defined in the Barrett-Lennard study) the process of change will be found to be the same in the schizophrenic person as in the normal.

Perhaps these brief descriptions are sufficient to indicate that the body of objective investigation stimulated by the practice and theory of client-centered therapy is continuing to grow and ramify.

THE MEANING OF RESEARCH FOR THE FUTURE

In concluding this chapter I would like to comment on the question "Where does this lead? To what end is all this research?"

Its major significance, it seems to me, is that a growing body of objectively verified knowledge of psychotherapy will bring about the gradual demise of "schools" of psychotherapy, including this one. As solid knowledge increases as to the conditions which facilitate therapeutic change, the nature of the therapeutic process, the conditions which block or inhibit therapy, the characteristic outcomes of therapy in terms of personality or behavioral change, then there will be less and less emphasis upon dogmatic and purely theoretical formulations. Differences of opinion, different procedures in therapy, different judgments as to outcome, will be put to empirical test rather than being simply a matter of debate or argument.

In medicine today we do not find a "penicillin school of treatment" versus some other school of treatment. There are differences of judgment and opinion, to be sure, but there is confidence that these will be resolved in the foreseeable future by carefully designed research. Just so I believe will psychotherapy turn increasingly to the facts rather than to dogma as an arbiter of differences.

Out of this should grow an increasingly effective, and continually changing psychotherapy which will neither have nor need any specific label. It will have incorporated whatever is factually verified from any and every therapeutic orientation.

Perhaps I should close here, but I would like to say one further word to those who may abhor research in such a delicately personal and intangible field as psychotherapy. They may feel that to subject such an intimate relationship to objective scrutiny is somehow to depersonalize it, to rob it of its most essential qualities, to reduce it to a cold system of facts. I would simply like to point out that to date this has not been its effect. Rather the contrary has been true. The more extensive the research the more it has become evident that the significant changes in the client have to do with very subtle and subjective experiences — inner choices, greater oneness within the whole person, a different feeling about one's self. And in the therapist some of the recent studies suggest that a warmly human and genuine therapist, interested only in understanding the moment-by-moment feelings of this person who is coming into being in the relationship with him, is the most effective therapist. Certainly there is nothing to indicate that the coldly intellectual analytical factually-minded therapist is effective. It seems to be one of the paradoxes of psychotherapy that to advance in our understanding of the field the individual must be willing to put his most passionate beliefs and firm convictions to the impersonal test of empirical research; but to be effective as a therapist, he must use this knowledge only to enrich and enlarge his subjective self, and must be that self, freely and without fear, in his relationship to his client.

References

1. Barrett-Lennard, G. T. Dimensions of the client's experience of his therapist associated with personality change. Unpublished doctoral dissertation, Univ. of Chicago, 1959.

2. Bergman, D. V. Counseling method and client responses. *J. Consult. Psychol.* 1951, *15*, 216–224.

3. Butler, J. M., and G. V. Haigh. Changes in the relation between self-concepts and ideal concepts consequent upon client-centered counseling. In C. R. Rogers and Rosalind F. Dymond (Eds.). *Psychotherapy and Personality Change.* University of Chicago Press, 1954, pp. 55–75.

4. Cartwright, Desmond S. Annotated bibliography of research and theory construction in client-centered therapy. *J. of Counsel. Psychol.* 1957, *4*, 82–100.

5. Raskin, N. J. An objective study of the locus-of-evaluation factor in psychotherapy. In W. Wolff, and J. A. Precker (Eds.). *Success in Psychotherapy.* New York: Grune & Stratton, 1952, Chap. 6.

6. Rogers, C. R. Changes in the maturity of behavior as related to therapy. In C. R. Rogers, and Rosalind F. Dymond (Eds.). *Psychotherapy and Personality Change.* University of Chicago Press, 1954, pp. 215–237.

7. Rogers, C. R. A process conception of psychotherapy. *Amer. Psychol.*, 1958, *13*, 142–149.

8. Rogers, C. R. and Dymond, R. F. (Eds.). *Psychotherapy and Personality Change.* University of Chicago Press, 1954, 447 p.

9. Seeman, J., and N. J. Raskin. Research perspectives in client centered therapy. In O. H. Mowrer (Ed.). *Psychotherapy: theory and research,* New York: Ronald, 1953, pp. 205–234.

10. Stephenson, W. *The Study of Behavior.* University of Chicago Press, 1953.

11. Thetford, William N., An objective measurement of frustration tolerance in evaluating psychotherapy. In W. Wolff, and J. A. Precker (Eds.). *Success in Psychotherapy,* New York: Grune & Stratton, 1952, Chapter 2.

PART VI

What Are the Implications for Living?

I have found the experience of therapy
to have meaningful and sometimes profound implications
for education, for interpersonal communication,
for family living, for the creative process.

13

Personal Thoughts on Teaching and Learning

This is the shortest chapter in the book but if my experience with it is any criterion, it is also the most explosive. It has a (to me) amusing history.

I had agreed, months in advance, to meet with a conference organized by Harvard University on "Classroom Approaches to Influencing Human Behavior." I was requested to put on a demonstration of "student-centered teaching" — teaching based upon therapeutic principles as I had been endeavoring to apply them in education. I felt that to use two hours with a sophisticated group to try help to them formulate their own purposes, and to respond to their feelings as they did so, would be highly artificial and unsatisfactory. I did not know what I would do or present.

At this juncture I took off for Mexico on one of our winter-quarter trips, did some painting, writing, and photography, and immersed myself in the writings of Søren Kierkegaard. I am sure that his honest willingness to call a spade a spade influenced me more than I realized.

As the time came near to return I had to face up to my obligation. I recalled that I had sometimes been able to initiate very meaningful class discussions by expressing some highly personal opinion of my own, and then endeavoring to understand and accept the often very divergent reactions and feelings of the students. This seemed a sensible way of handling my Harvard assignment.

So I sat down to write, as honestly as I could, what my experiences had been with teaching, *as this term is defined in the dictionaries, and likewise my experience with* learning. *I was far away from psychologists, educators, cautious colleagues. I simply put down what I felt, with assurance that if I had not got it correctly, the discussion would help to set me on the right track.*

I may have been naive, but I did not consider the material inflammatory. After all the conference members were knowledgeable, self-critical teachers, whose main common bond was an interest in the discussion method in the classroom.

I met with the conference, I presented my views as written out below, taking only a very few moments, and threw the meeting open for discussion. I was hoping for a response, but I did not expect the tumult which followed. Feelings ran high. It seemed I was threatening their jobs, I was obviously saying things I didn't mean, etc., etc. And occasionally a quiet voice of appreciation arose from a teacher who had felt these things but never dared to say them.

I daresay that not one member of the group remembered that this meeting was billed as a demonstration of student-centered teaching. But I hope that in looking back each realized that he had lived an experience of student-centered teaching. I refused to defend myself by replying to the questions and attacks which came from every quarter. I endeavored to accept and empathize with the indignation, the frustration, the criticisms which they felt. I pointed out that I had merely expressed some very personal views of my own. I had not asked nor expected others to agree. After much storm, members of the group began expressing, more and more frankly, their own significant feelings about teaching — often feelings divergent from mine, often feelings divergent from each other. It was a very thought-provoking session. I question whether any participant in that session has ever forgotten it.

The most meaningful comment came from one of the conference members the next morning as I was preparing to leave the city. All he said was, "You kept more people awake last night!"

I took no steps to have this small fragment published. My views on psychotherapy had already made me a "controversial figure" among psychologists and psychiatrists. I had no desire to add educators to the list. The statement was widely duplicated however by members of the conference and several years later two journals requested permission to publish it.

After this lengthy historical build-up, you may find the statement itself a let-down. Personally I have never felt it to be incendiary. It still expresses some of my deepest views in the field of education.

I WISH TO PRESENT some very brief remarks, in the hope that if they bring forth any reaction from you, I may get some new light on my own ideas.

I find it a very troubling thing to *think*, particularly when I think about my own experiences and try to extract from those experiences the meaning that seems genuinely inherent in them. At first such thinking is vary satisfying, because it seems to discover sense and pattern in a whole host of discrete events. But then it very often becomes dismaying, because I realize how ridiculous these thoughts, which have much value to me, would seem to most people. My impression is that if I try to find the meaning of my own experience it leads me, nearly always, in directions regarded as absurd.

So in the next three or four minutes, I will try to digest some of the meanings which have come to me from my classroom experience and the experience I have had in individual and group therapy. They are in no way intended as conclusions for some one else, or a guide to what others should do or be. They are the very tentative meanings, as of April 1952, which my experience has had for me, and some of the bothersome questions which their absurdity raises. I will put each idea or meaning in a separate lettered paragraph, not because they are in any particular logical order, but because each meaning is separately important to me.

a. I may as well start with this one in view of the purposes of this conference. *My experience has been that I cannot teach another person how to teach.* To attempt it is for me, in the long run, futile.

b. *It seems to me that anything that can be taught to another is relatively inconsequential, and has little or no significant influence on behavior.* That sounds so ridiculous I can't help but question it at the same time that I present it.

c. *I realize increasingly that I am only interested in learnings which signficantly influence behavior.* Quite possibly this is simply a personal idiosyncrasy.

d. *I have come to feel that the only learning which significantly influences behavior is self-discovered, self-appropriated learning.*

e. *Such self-discovered learning, truth that has been personally appropriated and assimilated in experience, cannot be directly communicated to another.* As soon as an individual tries to communicate such experience directly, often with a quite natural enthusiasm, it becomes teaching, and its results are inconsequential. It was some relief recently to discover that Søren Kierkegaard, the Danish philosopher, had found this too, in his own experience, and stated it very clearly a century ago. It made it seem less absurd.

f. As a consequence of the above, *I realize that I have lost interest in being a teacher.*

g. When I try to teach, as I do sometimes, I am appalled by the results, which seem a little more than inconsequential, because sometimes the teaching appears to succeed. When this happens I find that the results are damaging. It seems to cause the individual to distrust his own experience, and to stifle significant learning. *Hence I have come to feel that the outcomes of teaching are either unimportant or hurtful.*

h. When I look back at the results of my past teaching, the real results seem the same — either damage was done, or nothing significant occurred. This is frankly troubling.

i. As a consequence, *I realize that I am only interested in being a learner, preferably learning things that matter, that have some significant influence on my own behavior.*

j. *I find it very rewarding to learn,* in groups, in relationships with one person as in therapy, or by myself.

k. *I find that one of the best, but most difficult ways for me to learn is to drop my own defensiveness, at least temporarily, and to try to understand the way in which his experience seems and feels to the other person.*

l. *I find that another way of learning for me is to state my own uncertainties, to try to clarify my puzzlements, and thus get closer to the meaning that my experience actually seems to have.*

m. This whole train of experiencing, and the meanings that I have thus far discovered in it, seem to have launched me on a process which is both fascinating and at times a little frightening. *It seems to mean letting my experience carry me on, in a direction which appears to be forward, toward goals that I can but dimly define, as I try to understand at least the current meaning of that experience.* The sensation is that of floating with a complex stream of experience, with the fascinating possibility of trying to comprehend its ever changing complexity.

I am almost afraid I may seem to have gotten away from any discussion of learning, as well as teaching. Let me again introduce a practical note by saying that by themselves these interpretations of my own experience may sound queer and aberrant, but not particularly shocking. It is when I realize the *implications* that I shudder a bit at the distance I have come from the commonsense world that everyone knows is right. I can best illustrate that by saying that if the experiences of others had been the same as mine, and if they had discovered similar meanings in it, many consequences would be implied.

a. Such experience would imply that we would do away with teaching. People would get together if they wished to learn.

b. We would do away with examinations. They measure only the inconsequential type of learning.

c. The implication would be that we would do away with grades and credits for the same reason.

d. We would do away with degrees as a measure of competence partly for the same reason. Another reason is that a degree marks an end or a conclusion of something, and a learner is only interested in the continuing process of learning.

e. It would imply doing away with the exposition of conclusions, for we would realize that no one learns significantly from conclusions.

I think I had better stop there. I do not want to become too fantastic. I want to know primarily whether anything in my inward thinking as I have tried to describe it, speaks to anything in your experience of the classroom as you have lived it, and if so, what the meanings are that exist for you in *your* experience.

14

Significant Learning: In Therapy and in Education

Goddard College, at Plainfield, Vermont, is a small experimental college which in addition to its efforts on behalf of its students, frequently organizes conferences and workshops for educators, where they may deal with significant problems. I was asked to lead such a workshop in February 1958, on "The Implications of Psychotherapy for Education." Teachers and educational administrators from the eastern half of the country, and especially from the New England area, found their way through the thick snowdrifts to spend three concentrated days together.

I decided to try to reformulate my views on teaching and learning for this conference, hopefully in a way which would be less disturbing than the statement in the preceding chapter, yet without dodging the radical implications of a therapeutic approach. This paper is the result. For those who are familiar with Part II of this book the sections on "The Conditions of Learning in Psychotherapy" and "The Process of Learning in Therapy" will be redundant and may be skipped, since they are merely a restatement of the basic conditions for therapy, as described earlier.

To me this is the most satisfying formulation I have achieved of the meaning of the hypotheses of client-centered therapy in the field of education.

PRESENTED HERE IS A THESIS, a point of view, regarding the implications which psychotherapy has for education. It is a stand which I take tentatively, and with some hesitation. I have many unanswered questions about this thesis. But it has, I think, some clarity in it, and hence it may provide a starting point from which clear differences can emerge.

SIGNIFICANT LEARNING IN PSYCHOTHERAPY

Let me begin by saying that my long experience as a therapist convinces me that significant learning is facilitated in psychotherapy, and occurs in that relationship. By significant learning I mean learning which is more than an accumulation of facts. It is learning which makes a difference — in the individual's behavior, in the course of action he chooses in the future, in his attitudes and in his personality. It is a pervasive learning which is not just an accretion of knowledge, but which interpenetrates with every portion of his existence.

Now it is not only my subjective feeling that such learning takes place. This feeling is substantiated by research. In client-centered therapy, the orientation with which I am most familiar, and in which the most research has been done, we know that exposure to such therapy produces learnings, or changes, of these sorts:

The person comes see himself differently.

He accepts himself and his feelings more fully.

He becomes more self-confident and self-directing.

He becomes more the person he would like to be.

He becomes more flexible, less rigid, in his perceptions.

He adopts more realistic goals for himself.

He behaves in a more mature fashion.

He changes his maladjustive behaviors, even such a long-established one as chronic alcoholism.

He becomes more acceptant of others.

He becomes more open to the evidence, both to what is going on outside of himself, and to what is going on inside of himself.

He changes in his basic personality characteristics, in constructive ways.*

I think perhaps this is sufficient to indicate that these are learnings which are significant, which do make a difference.

Significant Learning in Education

I believe I am accurate in saying that educators too are interested in learnings which make a difference. Simple knowledge of facts has its value. To know who won the battle of Poltava, or when the umpteenth opus of Mozart was first performed, may win $64,000 or some other sum for the possessor of this information, but I believe educators in general are a little embarrassed by the assumption that the acquisition of such knowledge constitutes education. Speaking of this reminds me of a forceful statement made by a professor of agronomy in my freshman year in college. Whatever knowledge I gained in his course has departed completely, but I remember how, with World War I as his background, he was comparing factual knowledge with ammunition. He wound up his little discourse with the exhortation, "Don't be a damned ammunition wagon; be a rifle!" I believe most educators would share this sentiment that knowledge exists primarily for use.

To the extent then that educators are interested in learnings which are functional, which make a difference, which pervade the person and his actions, then they might well look to the field of psychotherapy for leads or ideas. Some adaptation for education of the learning process which takes place in psychotherapy seems like a promising possibility.

The Conditions of Learning in Psychotherapy

Let us then see what is involved, essentially, in making possible the learning which occurs in therapy. I would like to spell out, as clearly

* For evidence supporting these statements see references (7) and (9).

as I can, the conditions which seem to be present when this phenomenon occurs.

FACING A PROBLEM

The client is, first of all, up against a situation which he perceives as a serious and meaningful problem. It may be that he finds himself behaving in ways in which he cannot control, or he is overwhelmed by confusions and conflicts, or his marriage is going on the rocks, or he finds himself unhappy in his work. He is, in short, faced with a problem with which he has tried to cope, and found himself unsuccessful. He is therefore eager to learn, even though at the same time he is frightened that what he discovers in himself may be disturbing. Thus one of the conditions nearly always present is an uncertain and ambivalent desire to learn or to change, growing out of a perceived difficulty in meeting life.

What are the conditions which this individual meets when he comes to a therapist? I have recently formulated a theoretical picture of the necessary and sufficient conditions which the therapist provides, if constructive change or significant learning is to occur (8). This theory is currently being tested in several of its aspects by empirical research, but it must still be regarded as theory based upon clinical experience rather than proven fact. Let me describe briefly the conditions which it seems essential that the therapist should provide.

CONGRUENCE

If therapy is to occur, it seems necessary that the therapist be, in the relationship, a unified, or integrated, or congruent person. What I mean is that within the relationship he is exactly what he *is* — not a façade, or a role, or a pretense. I have used the term "congruence" to refer to this accurate matching of experience with awareness. It is when the therapist is fully and accurately aware of what he is experiencing at this moment in the relationship, that he is fully congruent. Unless this congruence is present to a considerable degree it is unlikely that significant learning can occur.

Though this concept of congruence is actually a complex one, I believe all of us recognize it in an intuitive and commonsense way in individuals with whom we deal. With one individual we recognize that he not only means exactly what he says, but that his deepest feelings also match what he is expressing. Thus whether he is angry or affectionate or ashamed or enthusiastic, we sense that he is the same at all levels — in what he is experiencing at an organismic level, in his awareness at the conscious level, and in his words and communications. We furthermore recognize that he is acceptant of his immediate feelings. We say of such a person that we know "exactly where he stands." We tend to feel comfortable and secure in such a relationship. With another person we recognize that what he is saying is almost certainly a front or a façade. We wonder what he *really* feels, what he is really experiencing, behind this façade. We may also wonder if *he* knows what he really feels, recognizing that he may be quite unaware of the feelings he is actually experiencing. With such a person we tend to be cautious and wary. It is not the kind of relationship in which defenses can be dropped or in which significant learning and change can occur.

Thus this second condition for therapy is that the therapist is characterized by a considerable degree of congruence in the relationship. He is freely, deeply, and acceptantly himself, with his actual experience of his feelings and reactions matched by an accurate awareness of these feelings and reactions as they occur and as they change.

Unconditional Positive Regard

A third condition is that the therapist experiences a warm caring for the client — a caring which is not possessive, which demands no personal gratification. It is an atmosphere which simply demonstrates "I care"; not "I care for you *if* you behave thus and so." Standal (11) has termed this attitude "unconditional positive regard," since it has no conditions of worth attached to it. I have often used the term "acceptance" to describe this aspect of the therapeutic climate. It involves as much feeling of acceptance for the client's expression of negative, "bad," painful, fearful, and abnormal feelings as for his expression of "good," positive, mature, confident and social feelings. It involves an acceptance of and a caring for the client as a *separate* person, with permission for him to have his own feelings and experiences, and to find his own meanings in them. To the degree that the therapist can provide this safety-creating climate of unconditional positive regard, significant learning is likely to take place.

AN EMPATHIC UNDERSTANDING

The fourth condition for therapy is that the therapist is experiencing an accurate, empathic understanding of the client's world as seen from the inside. To sense the client's private world as if it were your own, but without ever losing the "as if" quality — this is empathy, and this seems essential to therapy. To sense the client's anger, fear, or confusion as if it were your own, yet without your own anger, fear, or confusion getting bound up in it, is the condition we are endeavoring to describe. When the client's world is this clear to the therapist, and he moves about in it freely, then he can both communicate his understanding of what is clearly known to the client and can also voice meanings in the client's experience of which the client is scarcely aware. That such penetrating empathy is important for therapy is indicated by Fiedler's research in which items such as the following placed high in the description of relationships created by experienced therapists:

The therapist is well able to understand the patient's feelings.

The therapist is never in any doubt about what the patient means.

The therapist's remarks fit in just right with the patient's mood and content.

The therapist's tone of voice conveys the complete ability to share the patient's feelings. (3)

FIFTH CONDITION

A fifth condition for significant learning in therapy is that the client should experience or perceive something of the therapist's congruence, acceptance, and empathy. It is not enough that these conditions exist in the therapist. They must, to some degree, have been successfully communicated to the client.

THE PROCESS OF LEARNING IN THERAPY

It has been our experience that when these five conditions exist, a process of change inevitably occurs. The client's rigid perceptions of himself and of others loosen and become open to reality. The rigid ways in which he has construed the meaning of his experience are looked at, and he finds himself questioning many of the "facts" of his life, discovering that they are only "facts" because he has regarded them so. He discovers feelings of which he has been unaware, and experiences them, often vividly, in the therapeutic relationship. Thus he learns to be more open to all of his experience — the evidence within himself as well as the evidence without. He learns to *be* more of his experience — to be the feelings of which he has been frightened as well as the feelings he has regarded as more acceptable. He becomes a more fluid, changing, learning person.

The Mainspring of Change

In this process it is not necessary for the therapist to "motivate" the client or to supply the energy which brings about the change. Nor, in some sense, is the motivation supplied by the client, at least in any conscious way. Let us say rather that the motivation for learning and change springs from the self-actualizing tendency of life itself, the tendency for the organism to flow into all the differentiated channels of potential development, insofar as these are experienced as enhancing.

I could go on at very considerable length on this, but it is not my purpose to focus on the process of therapy and the learnings which take place, nor on the motivation for these learnings, but rather on the conditions which make them possible. So I will simply conclude this description of therapy by saying that it is a type of significant learning which takes place when five conditions are met:

When the client perceives himself as faced by a serious and meaningful problem;

When the therapist is a congruent person in the relationship, able to *be* the person he *is;*

When the therapist feels an unconditional positive regard for the client;

When the therapist experiences an accurate empathic understanding of the client's private world, and communicates this;

When the client to some degree experiences the therapist's congruence, acceptance, and empathy.

IMPLICATIONS FOR EDUCATION

What do these conditions mean if applied to education? Undoubtedly the teacher will be able to give a better answer than I out of his own experience, but I will at least suggest some of the implications.

CONTACT WITH PROBLEMS

In the first place it means that significant learning occurs more readily in relation to situations perceived as problems. I believe I have observed evidence to support this. In my own varying attempts to conduct courses and groups in ways consistent with my therapeutic experience, I have found such an approach more effective, I believe, in workshops than in regular courses, in extension courses than in campus courses. Individuals who come to workshops or extension courses are those who are in contact with problems which they recognize as problems. The student in the regular university course, and particularly in the required course, is apt to view the course as an experience in which he expects to remain passive or resentful or both, an experience which he certainly does not often see as relevant to his own problems.

Yet it has also been my experience that when a regular university class does perceive the course as an experience they can use to resolve problems which *are* of concern to them, the sense of release, and the thrust of forward movement is astonishing. And this is true of courses as diverse as Mathematics and Personality.

I believe the current situation in Russian education also supplies evidence on this point. When a whole nation perceives itself as being faced with the urgent problem of being behind — in agriculture, in industrial production, in scientific development, in weapons development — then an astonishing amount of significant learning takes place, of which the Sputniks are but one observable example.

So the first implication for education might well be that we permit the student, at any level, to be in real contact with the relevant problems of his existence, so that he perceives problems and issues which he wishes to resolve. I am quite aware that this implication, like the others I shall mention, runs sharply contrary to the current trends in our culture, but I shall comment on that later.

I believe it would be quite clear from my description of therapy that an overall implication for education would be that the task of the teacher is to create a facilitating classroom climate in which significant learning can take place. This general implication can be broken down into several sub-sections.

The Teacher's Real-ness

Learning will be facilitated, it would seem, if the teacher is congruent. This involves the teacher's being the person that he is, and being openly aware of the attitudes he holds. It means that he feels acceptant toward his own real feelings. Thus he becomes a real person in the relationship with his students. He can be enthusiastic about subjects he likes, and bored by topics he does not like. He can be angry, but he can also be sensitive or sympathetic. Because he accepts his feeling as *his* feelings, he has no need to impose them on his students, or to insist that they feel the same way. He is a *person*, not a faceless embodiment of a curricular requirement, or a sterile pipe through which knowledge is passed from one generation to the next.

I can suggest only one bit of evidence which might support this view. As I think back over a number of teachers who have facilitated my own learning, it seems to me each one has this quality of being a real person. I wonder if your memory is the same. If so, perhaps it is less important that a teacher cover the allotted amount of the curriculum, or use the most approved audio-visual devices, than that he be congruent, real, in his relation to his students.

Acceptance and Understanding

Another implication for the teacher is that significant learning may take place if the teacher can accept the student as he is, and can understand the feelings he possesses. Taking the third and fourth conditions of therapy as specified above, the teacher who can warmly accept, who can provide an unconditional positive regard, and who can empathize with the feelings of fear, anticipation, and discouragement which are involved in meeting new material, will have done a great deal toward setting the conditions for learning. Clark Moustakas, in his book, *The Teacher and the Child* (5), has given many excellent examples of individual and group situations from kindergarten to high school, in which the teacher has worked toward just this type of goal. It will perhaps disturb some that when the teacher holds such attitudes, when he is willing to be acceptant of feelings, it is not only attitudes toward school work itself which are expressed, but feelings about parents, feelings of hatred for brother or sister, feelings of concern about self — the whole gamut of attitudes. Do such feelings have a right to exist openly in a school setting? It is my thesis that they do. They are related to the person's becoming, to his effective learning and effective functioning, and to deal understandingly and acceptantly with such feelings has a definite relationship to the learning of long division or the geography of Pakistan.

PROVISION OF RESOURCES

This brings me to another implication which therapy holds for education. In therapy the resources for learning one's self lie within. There is very little data which the therapist can supply which will be of help since the data to be dealt with exist within the person. In education this is not true. There are many resources of knowledge, of techniques, of theory, which constitute raw material for use. It seems to me that what I have said about therapy suggests that these materials, these resources, be made available to the students, not forced upon them. Here a wide range of ingenuity and sensitivity is an asset.

I do not need to list the usual resources which come to mind — books, maps, workbooks, materials, recordings, work-space, tools, and the like. Let me focus for a moment on the way the teacher uses himself and his knowledge and experience as a resource. If the teacher holds the point of view I have been expressing then he would probably want to make himself available to his class in at least the following ways:

He would want to let them know of special experience and knowledge he has in the field, and to let them know they could call on this knowledge. Yet he would not want them to feel that they must use him in this way.

He would want them to know that his own way of thinking about the field, and of organizing it, was available to them, even in lecture form, if they wished. Yet again he would want this to be perceived as an offer, which could as readily be refused as accepted.

He would want to make himself known as a resource-finder. Whatever might be seriously wanted by an individual or by the whole group to promote their learning, he would be very willing to consider the possibilities of obtaining such a resource.

He would want the quality of his relationship to the group to be such that his feelings could be freely available to them, without being imposed on them or becoming a restrictive influence on them. He thus could share the excitements and enthusiasms of his own learnings, without insisting that the students follow in his footsteps; the feelings of disinterest, satisfaction, bafflement, or pleasure which he feels toward individual or group activities, without this becoming either a carrot or a stick for the student. His hope would be that he could say, simply for himself, "I don't like that," and that the student with equal freedom could say, "But I do."

Thus whatever the resource he supplies — a book, space to work, a new tool, an opportunity for observation of an industrial process, a lecture based on his own study, a picture, graph or map, his own emotional reactions — he would feel that these were, and would hope they would be perceived as, offerings to be used if they were useful to the student. He would not feel them to be guides, or expectations, or commands, or impositions or requirements. He would offer himself, and all the other resources he could discover, for use.

The Basic Motive

It should be clear from this that his basic reliance would be upon the self-actualizing tendency in his students. The hypothesis upon which he would build is that students who are in real contact with life problems wish to learn, want to grow, seek to find out, hope to master, desire to create. He would see his function as that of developing such a personal relationship with his students, and such a climate in his classroom, that these natural tendencies could come to their fruition.

SOME OMISSIONS

These I see as some of the things which are implied by a therapeutic viewpoint for the educational process. To make them a bit sharper, let me point out some of the things which are not implied.

I have not included lectures, talks, or expositions of subject matter which are imposed on the students. All of these procedures might be a part of the experience if they were desired, explicitly or implicitly, by the students. Yet even here, a teacher whose work was following through a hypothesis based on therapy would be quick to sense a shift in that desire. He might have been requested to lecture to the group (and to give a *requested* lecture is *very* different from the usual classroom experience), but if he detected a growing disinterest and boredom, he would respond to that, trying to understand the feeling which had arisen in the group, since his response to their feelings and attitudes would take precedence over his interest in expounding material.

I have not included any program of evaluation of the student's learnings in terms of external criteria. I have not, in other words, included examinations. I believe that the testing of the student's achievements in order to see if he meets some criterion held by the teacher, is directly contrary to the implications of therapy for significant learning. In therapy, the examinations are set by *life*. The client meets them, sometimes passing, sometimes failing. He finds that he can use the resources of the therapeutic relationship and his experience in it to organize himself so that he can meet life's tests more satisfyingly next time. I see this as the paradigm for education also. Let me try to spell out a fantasy of what it would mean.

In such an education, the requirements for many life situations would be a part of the resources the teacher provides. The student would have available the knowledge that he cannot enter engineering school without so much math; that he cannot get a job in X corporation unless he has a college diploma; that he cannot become a psychologist without doing an independent doctoral research; that he cannot be a doctor without knowledge of chemistry; that he cannot even drive a car without passing an examination on rules of the road. These are requirements set, not by the teacher, but by life. The teacher is there to provide the resources which the student can use to learn so as to be able to meet these tests.

There would be other in-school evaluations of similar sort. The student might well be faced with the fact that he cannot join the Math Club until he makes a certain score on a standardized mathematics test; that he cannot develop his camera film until he has shown an adequate knowledge of chemistry and lab techniques; that he cannot join the special literature section until he has shown evidence of both wide reading and creative writing. The natural place of evaluation in life is as a ticket of entrance, not as a club over the recalcitrant. Our experience in therapy would suggest that it should be the same way in the school. It would leave the student as a self-respecting, self-motivated person, free to choose whether he wished to put forth the effort to gain these tickets of entrance. It would thus refrain from forcing him into conformity, from sacrificing his creativity, and from causing him to live his life in terms of the standards of others.

I am quite aware that the two elements of which I have just been speaking — the lectures and expositions imposed by the teacher on the group, and the evaluation of the individual by the teacher, consitute the two major ingredients of current education. So when I say that experience in psychotherapy would suggest that they both be omitted, it should be quite clear that the implications of psychotherapy for education are startling indeed.

Probable Outcomes

If we are to consider such drastic changes as I have outlined, what would be the results which would justify them? There have been some research investigations of the outcomes of a student-centered type of teaching (1, 2, 4), though these studies are far from adequate. For one thing, the situations studied vary greatly in the extent to which they meet the conditions I have described. Most of them have extended only over a period of a few months, though one recent study with lower class children extended over a full year (4). Some involve the use of adequate controls, some do not.

I think we may say that these studies indicate that in classroom situations which at least attempt to approximate the climate I have described, the findings are as follows: Factual and curricular learning is roughly equal to the learning in conventional classes. Some studies report slightly more, some slightly less. The student-centered group shows gains significantly greater than the conventional class in personal adjustment, in self-initiated extra-curricular learning, in creativity, in self-responsibility.

I have come to realize, as I have considered these studies, and puzzled over the design of better studies which should be more informative and conclusive, that findings from such research will never answer our questions. For all such findings must be evaluated in terms of the goals we have for education. If we value primarily the learning of knowledge, then we may discard the conditions I have described as useless, since there is no evidence that they lead to a greater rate or amount of factual knowledge. We may then favor such measures as the one which I understand is advocated by a number of members of Congress — the setting up of a training school for scientists, modeled upon the military academies. But if we value creativity, if we deplore the fact that all of our germinal ideas in atomic physics, in psychology, and in other sciences have been borrowed from Europe, then we may wish to give a trial to ways of facilitating learning which give more promise of freeing the mind. If we value independence, if we are disturbed by the growing conformity of knowledge, of values, of attitudes, which our present system induces, then we may wish to set up conditions of learning which make for uniqueness, for self-direction, and for self-initiated learning.

Some Concluding Issues

I have tried to sketch the kind of education which would be implied by what we have learned in the field of psychotherapy. I have endeavored to suggest very briefly what it would mean if the central focus of the teacher's effort were to develop a relationship, an atmosphere, which was conducive to self-motivated, self-actualizing, significant learning. But this is a direction which leads sharply away from current educational practices and educational trends. Let me mention a few of the very diverse issues and questions which need to be faced if we are to think constructively about such an approach.

In the first place, how do we conceive the goals of education? The approach I have outlined has, I believe, advantages for achieving certain goals, but not for achieving others. We need to be clear as to the way we see the purposes of education.

What are the actual outcomes of the kind of education I have described? We need a great deal more of rigorous, hard-headed research to know the actual results of this kind of education as compared with conventional education. Then we can choose on the basis of the facts.

Even if we were to try such an approach to the facilitation of learning, there are many difficult issues. Could we possibly permit students to come in contact with real issues? Our whole culture — through custom, through the law, through the efforts of labor unions and management, through the attitudes of parents and teachers — is deeply committed to keeping young people away from any touch with real problems. They are not to work, they should not carry responsibility, they have no business in civic or political problems, they have no place in international concerns, they simply should be guarded from any direct contact with the real problems of individual and group living. They are not expected to help about the home, to earn a living, to contribute to science, to deal with moral issues. This is a deep seated trend which has lasted for more than a generation. Could it possibly be reversed?

Another issue is whether we could permit knowledge to be organized in and by the individual, or whether it is to be organized *for* the individual. Here teachers and educators line up with parents and national leaders to insist that the pupil must be guided. He must be inducted into knowledge we have organized for him. He cannot be trusted to organize knowledge in functional terms for himself. As Herbert Hoover says of high school students, "You simply cannot expect kids of those ages to determine the sort of education they need unless they have some guidance."* This seems so obvious to most people that even to question it is to seem somewhat unbalanced. Even a chancellor of a university questions whether freedom is really necessary in education, saying that perhaps we have overestimated its value. He says the Russians have advanced mightily in science without it, and implies that we should learn from them.

Still another issue is whether we would wish to oppose the strong current trend toward education as drill in factual knowledge. All must learn the same facts in the same way. Admiral Rickover states it as his belief that "in some fashion we must devise a way to introduce uniform standards into American education. . . . For the first time, parents would have a real yardstick to measure their schools. If the local school continued to teach such pleasant subjects as 'life adjustment' . . . instead of French and physics, its diploma would be, for all the world to see, inferior."† This is a statement of a very prevalent view. Even such a friend of forward-looking views in education as Max Lerner says at one point, "All that a school can ever hope to do is to equip the student with tools which he can later use to become an educated man" (5, p. 741). It is quite clear that he despairs of significant learning taking place in our school system, and feels that it must take place outside. All the school can do is to pound in the tools.

One of the most painless ways of inculcating such factual tool knowledge is the "teaching machine" being devised by B. F. Skinner and his associates (10). This group is demonstrating that the teacher is an outmoded and ineffective instrument for teaching arithmetic, trigonometry, French, literary appreciation, geography, or other factual subjects. There is simply no doubt in my mind that these teaching machines, providing immediate rewards for "right" answers, will be further developed, and will come into wide use. Here is a new contribution from the field of the behavioral sciences with which we must come to terms. Does it take the place of the approach I have described, or is it supplemental to it? Here is one of the problems we must consider as we face toward the future.

* *Time*, December 2, 1957.
† *Ibid.*

I hope that by posing these issues, I have made it clear that the double-barreled question of what constitutes significant learning, and how it is to be achieved, poses deep and serious problems for all of us. It is not a time when timid answers will suffice. I have tried to give a definition of significant learning as it appears in psychotherapy, and a description of the conditions which facilitate such learning. I have tried to indicate some implications of these conditions for education. I have, in other words, proposed one answer to these questions. Perhaps we can use what I have said, against the twin backdrops of current public opinion and current knowledge in the behavioral sciences, as a start for discovering some fresh answers of our own.

References

1. Faw, Volney. A psychotherapeutic method of teaching psychology. *Amer. Psychol. 4*: 104–09, 1949.

2. Faw, Volney. "Evaluation of student-centered teaching." Unpublished manuscript, 1954.

3. Fiedler, F. E. A comparison of therapeutic relationships in psychoanalytic, non-directive and Adlerian therapy. *J. Consult. Psychol.* 1950, *14*, 436–45.

4. Jackson, John H. The relationship between psychological climate and the quality of learning outcomes among lower-status pupils. Unpublished Ph.D. thesis, University of Chicago, 1957.

5. Lerner, Max. *America as a Civilization.* New York: Simon & Schuster, 1957.

6. Moustakas, Clark. *The Teacher and the Child.* New York: McGraw-Hill, 1956.

7. Rogers, C. R. *Client-Centered Therapy.* Boston: Houghton Mifflin Co., 1951.

8. Rogers, C. R. The necessary and sufficient conditions of therapeutic personality change. *J. Consult. Psychol.* 1957, 21, 95–103.

9. Rogers, C. R., and R. Dymond, (Eds.). *Psychotherapy and Personality Change.* University of Chicago Press, 1954.

10. Skinner, B. F. The science of learning and the art of teaching. *Harvard Educational Review* 1954, *24*, 86–97.

11. Standal, Stanley. The need for positive regard: A contribution to client-centered theory. Unpublished Ph.D. thesis, University of Chicago, 1954.

15

Student-Centered Teaching as Experienced by a Participant

It will have been evident earlier in this volume that I cannot be content simply to give my view of psychotherapy: I regard it as essential to give the client's perception of the experience also, since this is indeed the raw material from which I have formulated my own views. In the same way I found I could not be content simply to formulate my views of what education is when it is built upon the learnings from psychotherapy: I wanted to give the student's perception of such education also.

To this end I considered the various reports and "reaction sheets" which I have assembled from students in different courses over the years. Excerpts from these would have fulfilled my purpose. In the end, however, I chose to use two documents written by Dr. Samuel Tenenbaum, the first immediately after his participation in a course of mine, the second a letter to me one year later. I am deeply grateful to him for his permission to use these personal statements. I would like to place them in context for the reader.

In the summer of 1958 I was invited to teach a four-week course at Brandeis University. My recollection is that the title was "The

Process of Personality Change." I had no great expectations for the course. It was to be one of several courses which the students were taking, meeting for three two-hour sessions per week, rather than the concentrated workshop pattern which I prefer. I learned in advance that the group was to be unusually heterogeneous — teachers, doctoral candidates in psychology, counselors, several priests, at least one from a foreign country, psychotherapists in private practice, school psychologists. The group was, on the average, more mature and experienced than would ordinarily be found in a university course. I felt very relaxed about the whole thing. I would do what I could to help make this a meaningful experience for us all, but I doubted that it could have the impact of, for example, the workshops on counseling which I had conducted.

Perhaps it was because I had very modest expectations of the group and of myself, that it went so well. I would without doubt class it as among the most satisfying of my attempts to facilitate learning in courses or workshops. This should be borne in mind in reading Dr. Tenenbaum's material.

I would like to digress for a moment here to say that I feel far more assurance in confronting a new client in therapy than I do in confronting a new group. I feel I have a sufficient grasp of the conditions of therapy so that I have a reasonable confidence as to the process which will ensue. But with groups I have much less confidence. Sometimes when I have had every reason to suppose a course would go well, the vital, self-initiated, self-directed learning has simply not occurred to any great degree. At other times when I have been dubious, it has gone extremely well. To me this means that our formulation of the process of facilitating learning in education is not nearly as accurate or complete as our formulations regarding the therapeutic process.

But to return to the Brandeis summer course. It was clearly a highly significant experience for almost all of the participants, as evident in their reports on the course. I was particularly interested in the report by Dr. Tenenbaum, written as much for his colleagues as for me. Here was a mature scholar, not an impressionable young student. Here was a sophisticated educator, who already had to his credit a published biography of William H. Kilpatrick, the philoso-

pher of education. Hence his perceptions of the experience seemed unusually valuable.

I would not want it to be understood that I shared all of Dr. Tenenbaum's perceptions. Portions of the experience I perceived quite differently, but this is what made his observations so helpful. I felt particularly concerned that it seemed to him so much a "Rogers" approach, that it was simply my person and idiosyncrasies which made the experience what it was.

For this reason I was delighted to get a long letter from him a year later, reporting his own experience in teaching. This confirmed what I have learned from a wide variety of individuals, that it is not simply the personality of a specific teacher which makes this a dynamic learning experience, but the operation of certain principles which may be utilized by any "facilitator" who holds the appropriate attitudes.

I believe the two accounts by Dr. Tenenbaum will make it clear why teachers who have experienced the kind of group learning which is described can never return to more stereotyped ways of education. In spite of frustration and occasional failure, one keeps trying to discover, with each new group, the conditions which will unleash this vital learning experience.

Carl R. Rogers and Non-Directive Teaching

by Samuel Tenenbaum, Ph.D.

AS ONE INTERESTED in education, I have participated in a classroom methodology that is so unique and so special that I feel impelled to share the experience. The technique, it seems to me, is so radically different from the customary and the accepted, so undermining of the old, that it should be known more widely. As good a description of the process as any — I suppose the one that Carl R.

Rogers, the instructor, himself would be inclined to use — would be "non-directive" teaching.

I had some notion what that term meant, but frankly I was not prepared for anything that proved so overwhelming. It is not that I am convention-bound. My strongest educational influences stem from William Heard Kilpatrick and John Dewey, and anyone who has even the slightest acquaintance with their thinking would know that it does not smack of the narrow or the provincial. But this method which I saw Dr. Rogers carry out in a course which he gave at Brandeis University was so unusual, something I could not believe possible, unless I was part of the experience. I hope I shall manage to describe the method in a way to give you some inkling of the feelings, the emotions, the warmth and the enthusiasms that the method engendered.

The course was altogther unstructured; and it was exactly that. At no moment did anyone know, not even the instructor, what the next moment would bring forth in the classroom, what subject would come up for discussion, what questions would be raised, what personal needs, feelings and emotions aired. This atmosphere of non-structured freedom — as free as human beings could allow each other to be — was set by Dr. Rogers himself. In a friendly, relaxed way, he sat down with the students (about 25 in number) around a large table and said it would be nice if we stated our purpose and introduced ourselves. There ensued a strained silence; no one spoke up. Finally, to break it, one student timidly raised his hand and spoke his piece. Another uncomfortable silence, and then another upraised hand. Thereafter, the hands rose more rapidly. At no time did the instructor urge any student to speak.

UNSTRUCTURED APPROACH

Afterwards, he informed the class that he had brought with him quantities of materials — reprints, brochures, articles, books; he handed out a bibliography of recommended reading. At no time did he indicate that he expected students to read or do anything else. As I recall, he made only one request. Would some student volunteer to set up this material in a special room which had been reserved for students of the course? Two students promptly volunteered. He also said he had with him recorded tapes of therapeutic sessions and also reels of motion pictures. This created a flurry of excitement, and students asked whether they could be heard and seen and Dr. Rogers answered yes. The class then decided how it could be done best. Students volunteered to run tape recorders, find a movie projector; for the most part this too was student initiated and arranged.

Thereafter followed four hard, frustrating sessions. During this period, the class didn't seem to get anywhere. Students spoke at random, saying whatever came into their heads. It all seemed chaotic, aimless, a waste of time. A student would bring up some aspect of Rogers' philosophy; and the next student, completely disregarding the first, would take the group away in another direction; and a third, completely disregarding the first two, would start fresh on something else altogether. At times there were some faint efforts at a cohesive discussion, but for the most part the classroom proceedings seemed to lack continuity and direction. The instructor received every contribution with attention and regard. He did not find any student's contribution in order or out of order.

The class was not prepared for such a totally unstructured approach. They did not know how to proceed. In their perplexity and frustration, they demanded that the teacher play the role assigned to him by custom and tradition; that he set forth for us in authoritative language what was right and wrong, what was good and bad. Had they not come from far distances to learn from the oracle himself? Were they not fortunate? Were they not about to be initiated in the right rituals and practices by the great man himself, the founder of the movement that bears his name? The notebooks were poised for the climactic moment when the oracle would give forth, but mostly they remained untouched.

Queerly enough, from the outset, even in their anger, the members of the group felt joined together, and outside the classroom, there was an excitement and a ferment, for even in their frustration, they had communicated as never before in any classroom, and probably never before in quite the way they had. The class was bound together by a common, unique experience. In the Rogers class, they had spoken their minds; the words did not come from a book, nor were they the reflection of the instructor's thinking, nor that of any other authority. The ideas, emotions and feelings came from themselves; and this was the releasing and the exciting process.

In this atmosphere of freedom, something for which they had not bargained and for which they were not prepared, the students spoke up as students seldom do. During this period, the instructor took many blows; and it seemed to me that many times he appeared to be shaken; and although he was the source of our irritation, we had, strange as it may seem, a great affection for him, for it did not seem right to be angry with a man who was so sympathetic, so sensitive to the feelings and ideas of others. We all felt that what was involved was some slight misunderstanding, which once understood and remedied would make everything right again. But our instructor, gentle enough on the surface, had a "whim of steel." He didn't seem to understand; and if he did, he was obstinate and obdurate; he refused to come around. Thus did this tug-of-war continue. We all looked to Rogers and Rogers looked to us. One student, amid general approbation, observed: "We are Rogers-centered, not student-centered. We have come to learn from Rogers."

ENCOURAGING THINKING

Another student had discovered that Rogers had been influenced by Kilpatrick and Dewey, and using this idea as a springboard, he said he thought he perceived what Rogers was trying to get at. He thought Rogers wanted students to think independently, creatively; he wanted students to become deeply involved with their very persons, their very selves, hoping that this might lead to the "reconstruction" of the person — in the Dewey sense of the term — the person's outlook, attitudes, values, behavior. This would be a true reconstruction of experience; it would be learning in a real sense. Certainly, he didn't want the course to end in an examination based on textbooks and lectures, followed by the traditional end-term grade, which generally means completion and forgetting.* Rogers had expressed the belief almost from the outset of the course that no one can teach anyone else anything. But thinking, this student insisted, begins at the fork in the road, the famed dilemma set up by Dewey. As we reach the fork in the road, we do not know which road to take if we are to reach our destination; and then we begin to examine the situation. Thinking starts at that point.

Kilpatrick also sought original thinking from his students and also rejected a regurgitant textbook kind of learning, but he presented crucial problems for discussion, and these problems aroused a great deal of interest, and they also created vast changes in the person. Why can't committees of students or individual students get up such problems for discussion?† Rogers listened sympathetically and said, "I see you feel strongly about this?" That disposed of that. If I recall correctly, the next student who spoke completely disregarded what had been suggested and started afresh on another topic, quite in conformity with the custom set by the class.

Spasmodically, through the session, students referred favorably to the foregoing suggestion, and they began to demand more insistently that Rogers assume the traditional role of a teacher. At this point, the blows were coming Rogers' way rather frequently and strongly and I thought I saw him bend somewhat before them. (Privately, he denied he was so affected.) During one session, a student made the suggestion that he lecture one hour and that we have a class discus-

* It should be noted that Dr. Rogers neither agreed nor disagreed. It was not his habit to respond to students' contributions unless a remark was directed specifically to him; and even then he might choose not to answer. His main object, it seemed to me, was to follow students' contributions intelligently and sympathetically.

† One student compiled such a list, had it mimeographed, distributed it, and for practical purposes that was the end of that.

In this connection, another illustration may be in order. At the first session, Rogers brought to class tape recordings of therapeutic sessions. He explained that he was not comfortable in a teacher's role and he came "loaded," and the recordings served as a sort of security. One student continually insisted that he play the recordings, and after considerable pressure from the class, he did so, but he complied reluctantly; and all told, despite the pressure, he did not play them for more than an hour in all the sessions. Apparently, Rogers preferred the students to make real live recordings rather than listen to those which could only interest them in an academic way.

sion the next. This one suggestion seemed to fit into his plans. He said he had with him an unpublished paper. He warned us that it was available and we could read it by ourselves. But the student said it would not be the same. The person, the author, would be out of it, the stress, the inflection, the emotion, those nuances which give value and meaning to words. Rogers then asked the students if that was what they wanted. They said yes. He read for over an hour. After the vivid and acrimonious exchanges to which we had become accustomed, this was certainly a letdown, dull and soporific to the extreme. This experience squelched all further demands for lecturing. In one of the moments when he apologized for this episode ("It's better, more excusable, when students demand it."), he said: "You asked me to lecture. It is true I am a resource, but what sense would there be in my lecturing? I have brought a great quantity of material, reprints of any number of lectures, articles, books, tape recordings, movies."

By the fifth session, something definite had happened; there was no mistaking that. Students spoke to one another; they by-passed Rogers. Students asked to be heard and wanted to be heard, and what before was a halting, stammering, self-conscious group became an interacting group, a brand new cohesive unit, carrying on in a unique way; and from them came discussion and thinking such as no other group but this could repeat or duplicate. The instructor also joined in, but his role, more important than any in the group, somehow became merged with the group; the group was important, the center, the base of operation, not the instructor.

What caused it? I can only conjecture as to the reason. I believe that what happened was this: For four sessions students refused to believe that the instructor would refuse to play the traditional role. They still believed that he would set the tasks; that he would be the center of whatever happened and that he would manipulate the group. It took the class four sessions to realize that they were wrong; that he came to them with nothing outside of himself, outside of his own person; that if they really wanted something to happen, it was they who had to provide the content — an uncomfortable, challenging situation indeed. It was they who had to speak up, with all the risks that that entailed. As part of the process, they shared, they took exception, they agreed, they disagreed. At any rate, their persons, their deepest selves were involved; and from this situation, this special, unique group, this new creation was born.

Importance of Acceptance

As you may know, Rogers believes that if a person is accepted, fully accepted, and in this acceptance there is no judgment, only compassion and sympathy, the individual is able to come to grips with himself, to develop the courage to give up his defenses and face his true self. I saw this process work. Amid the early efforts to communicate, to find a *modus vivendi*, there had been in the group tentative exchanges of feelings, emotions and ideas; but after the fourth session, and progressively thereafter, this group, haphazardly thrown together, became close to one another and their true selves appeared. As they interacted, there were moments of insight and revelation and understanding that were almost awesome in nature; they were what, I believe, Rogers would describe as "moments of therapy," those pregnant moments when you see a human soul revealed before you, in all its breathless wonder; and then a silence, almost like reverence, would overtake the class. And each member of the class became enveloped with a warmth and a loveliness that border on the mystic. I for one, and I am quite sure the others also, never had an experience quite like this. It was learning and therapy; and by therapy I do not mean illness, but what might be characterized by a healthy change in the person, an increase in his flexibility, his openness, his willingness to listen. In the process, we all felt elevated, freer, more accepting of ourselves and others, more open to new ideas, trying hard to understand and accept.

This is not a perfect world, and there was evidence of hostility as members differed. Somehow in this setting every blow was softened, as if the sharp edges had been removed; if undeserved, students would go off to something else; and the blow was somehow lost. In my own case, even those students who originally irritated me, with further acquaintance I began to accept and respect; and the thought occurred to me as I tried to understand what was happening: Once you come close to a person, perceive his thoughts, his emotions, his feelings, he becomes not only understandable but good and desirable. Some of the more aggressive ones spoke more than they should, more than their right share, but the group itself, by its own being, not by setting rules, eventually made its authority felt; and unless a person was very sick or insensitive, members more or less, in this respect, conformed to what was expected of them. The problem — the hostile, the dominant, the neurotic — was not too acute; and yet if measured in a formal way, with a stop watch, at no time was a session free of aimless talk and waste of time. But yet as I watched the process, the idea persisted that perhaps this waste of time may be necessary; it may very well be that that is the way man learns best; for certainly, as I look back at the whole experience, I am fairly certain that it would have been impossible to learn as much or as well or as thoroughly in the traditional classroom setting. If we accept Dewey's definition of education as the reconstruction of experience, what better way can a person learn than by becoming involved with his whole self, his very person, his root drives, emotions, attitudes and values? No series of facts or arguments, no matter how logically or brilliantly arranged, can even faintly compare with that sort of thing.

In the course of this process, I saw hard, inflexible, dogmatic persons, in the brief period of several weeks, change in front of my eyes and become sympathetic, understanding and to a marked degree non-judgmental. I saw neurotic, compulsive persons ease up and become more accepting of themselves and others. In one instance, a student who particularly impressed me by his change, told me when I mentioned this: "It is true. I feel less rigid, more open to the world. And I like myself better for it. I don't believe I ever learned so much anywhere." I saw shy persons become less shy and aggressive persons more sensitive and moderate.

One might say that this appears to be essentially an emotional process. But that I believe would be altogether inaccurate in describing it. There was a great deal of intellectual content, but the intellectual content was meaningful and crucial to the person, in a sense that it meant a great deal to him as a person. In fact, one student brought up this very question. "Should we be concerned," he asked, "only with the emotions? Has the intellect no play?" It was my turn to ask, "Is there any student who has read as much or thought as much for any other course?"

The answer was obvious. We had spent hours and hours reading; the room reserved for us had occupants until 10 o'clock at night, and then many left only because the university guards wanted to close the building. Students listened to recordings; they saw motion pictures; but best of all, they talked and talked and talked. In the traditional course, the instructor lectures and indicates what is to be read and learned; students dutifully record all this in their note-books, take an examination and feel good or bad, depending on the outcome; but in nearly all cases it is a complete experience, with a sense of finality; the laws of forgetting begin to operate rapidly and inexorably. In the Rogers course, students read and thought inside and outside the class; it was they who chose from this reading and thinking what was meaningful to them, not the instructor.

This non-directive kind of teaching, I should point out, was not 100 per cent successful. There were three or four students who found the whole idea distasteful. Even at the end of the course, although nearly all became enthusiastic, one student to my knowledge, was intensely negative in his feelings; another was highly critical. These wanted the instructor to provide them with a rounded-out intellectual piece of merchandise which they could commit to memory and then give back on an examination. They would then have the assurance that they had learned what they should. As one said, "If I had to make a report as to what I learned in this course, what could I say?" Admittedly, it would be much more difficult than in a traditional course, if not impossible.

The Rogers method was free and flowing and open and permissive. A student would start an interesting discussion; it would be taken up by a second; but a third student might take us away in another direction, bringing up a personal matter of no interest to the class; and we would all feel frustrated. But this was like life, flowing on like a river, seemingly futile, with never the same water there, flowing on, with no one knowing what would happen the next moment. But in this there was an expectancy, an alertness, an aliveness; it seemed to me as near a smear of life as one could get in a classroom. For the authoritarian person, who puts his faith in neatly piled up facts, this method I believe can be threatening, for here he gets no reassurance, only an openness, a flowing, no closure.

A NEW METHODOLOGY

I believe that a great deal of the stir and the ferment that characterized the class was due to this lack of closure. In the lunch room, one could recognize Rogers' students by their animated discussions, by their desire to be together; and sometimes, since there was no table large enough, they would sit two and three tiers deep; and they would eat with plates on their laps. As Rogers himself points out, there is no finality in the process. He himself never summarizes (against every conventional law of teaching). The issues are left unresolved; the problems raised in class are always in a state of flux, on-going. In their need to know, to come to some agreement, students gather together, wanting understanding, seeking closure. Even in the matter of grades, there is no closure. A grade means an end; but Dr. Rogers does not give the grade; it is the student who suggests the grade; and since he does so, even this sign of completion is left unresolved, without an end, unclosed. Also, since the course is unstructured, each has staked his person in the course; he has spoken, not with the textbook as the gauge, but with his person, and thus as a self he has communicated with others, and because of this, in contradistinction to the impersonal subject matter that comprises the normal course, there develops this closeness and warmth.

To describe the many gracious acts that occurred might convey some idea of this feeling of closeness. One student invited the class to her home for a cookout. Another student, a priest from Spain, was so taken with the group that he talked of starting a publication to keep track of what was happening to the group members after they disbanded. A group interested in student counseling met on its own. A member arranged for the class to visit a mental hospital for children and adults; also he arranged for us to see the experimental work being done with psychotic patients by Dr. Lindsley. Class members brought in tape recordings and printed matter to add to the library material set aside for our use. In every way the spirit of good-will and friendliness was manifest to an extent that happens only in rare and isolated instances. In the many, many courses I have taken I have not seen the like. In this connection, it should be pointed out that the members comprised a group that had been haphazardly thrown together; they had come from many backgrounds and they included a wide age range.

I believe that what has been described above is truly a creative addition to classroom methodology; it is radically different from the old. That it has the capacity to move people, to make them freer, more open-minded, more flexible, I have no doubt. I myself witnessed the power of this method. I believe that non-directive teaching has profound implications which even those who accept this point of view cannot at present fully fathom. Its importance, I believe, goes beyond the classroom and extends to every area where human beings communicate and try to live with one another.

More specifically, as a classroom methodology, it warrants the widest discussion, inquiry and experimentation. It has the possibility of opening up a whole new dimension of thinking, fresh and original, for in its approach, in its practice, in its philosophy it differs so fundamentally from the old. It seems to me this approach ought to be tried out in every area of learning — elementary, high school, college, wherever human beings gather to learn and improve on the old. At this stage we should not be overly concerned about its limitations and inadequacies, since the method has not been refined and we do not know as much about it as we ought. As a new technique, it starts off with a handicap. We are loath to give up the old. The old is bolstered by tradition, authority and respectability; and we ourselves are its product. If we view education, however, as the reconstruction of experience, does not this presume that the individual must do his own reconstructing? He must do it himself, through the reorganization of his deepest self, his values, his attitudes, his very person. What better method is there to engross the individual; to bring him, his ideas, his feelings into communication with others; to break down the barriers that create isolation in a world where for his own mental safety and health, man has to learn to be part of mankind?

A Personal Teaching Experience

(as reported to Dr. Rogers one year later)

by

Samuel Tenenbaum, Ph.D.

I FEEL IMPELLED to write to you about my first experience in teaching after being exposed to your thinking and influence. You may or may not know I had a phobia about teaching. Since my work with you, I began to perceive more clearly where the difficulty lay. It was mostly in my concept of the role I had to play as a teacher — the motivator, director and the production chief of a performance. I always feared being "hung up" in the classroom — I believe it's your expression and I have come to like it — the class listless, uninterested, not responding, and my yammering and yammering, until I lost poise, the sentences not forming, coming out artificially, and the time moving slowly, slowly, ever more slowly. This was the horror I imagined. I suppose pieces of this happen to every teacher, but I would put them all together, and I would approach the class with foreboding, not at ease, not truly myself.

And now comes my experience. I was asked to give two summer courses for the Graduate School of Education of Yeshiva University, but I had a perfect alibi. I was going to Europe and I couldn't. Wouldn't I give an interim course, a concentrated course of 14 sessions during the month of June; and this would not interfere with the trip? I had no excuse and I accepted — because I no longer wanted to dodge the situation and more, also, because I was determined once and for all to face it. If I didn't like to teach (I haven't taught for nearly ten years), I would learn something. And if I did, I would also learn something. And if I had to suffer, it was best this way, since the course was concentrated and the time element was short.

You know that I have been strongly influenced in my thinking about education by Kilpatrick and Dewey. But now I had another powerful ingredient — you. When I first met my class, I did something I never did before. I was frank about my feelings. Instead of feeling that a teacher should know and students were there to be taught, I admitted weaknesses, doubts, dilemmas, and NOT KNOWING. Since I sort of dethroned my role as a teacher to the class and myself, my more natural self came out more freely and I found myself talking easily and even creatively. By "creatively" I mean ideas came to me as I spoke, brand new ideas which I felt were good.

Another important difference: It is true that since I was influenced by the Kilpatrick methodology I always welcomed the widest discussion, but I now know, I still wanted and expected my students to know the text and the lecture material set out for them. Even worse, I now know that although I welcomed discussion, I wanted, above all things, that, after all was said and done, the final conclusions of the class to come out according to my way of thinking. Hence none of the discussions were real discussions, in the sense that it was open and free and inquiring; none of the questions were real questions, in the sense that they sought to evoke thinking; all of them were loaded, in the sense that I had pretty definite convictions about what I thought were good answers and at times right answers. Hence, I came to the class with subject matter and my students were really instruments by which situations were manipulated to produce the inclusion of what I regarded as desirable subject matter.

In this last course, I didn't have the courage to discard all subject matter, but this time I really listened to my students; I gave them understanding and sympathy. Although I would spend hours and hours preparing for each session, I found that not once did I refer to a note from the voluminous material with which I entered the room. I allowed students free rein, not holding anyone down to any set course, and I permitted the widest diversion; and I followed wherever the students led.

I remember discussing this with a prominent educator and he said, in what I thought was a disappointed and disapproving tone: "You insist, of course, on good thinking." I quoted William James, who in effect said that man is a speck of reason in an ocean of emotion. I told him that I was more interested in what I would call a "third dimension," the feeling part of the students.

I cannot say I followed you all the way, Dr. Rogers, since I would express opinions and at times, unfortunately, lecture; and that I believe is bad, since students, once authoritative opinions are expressed, tend not to think, but to try to guess what is in the instructor's head and provide him with what he might like, so as to find favor in his eyes. If I had to do it over again, I would have less of that. But I did try and I believe I succeeded in large measure to give to each student a sense of dignity, respect and acceptance; farthest from my mind was to check on them or evaluate and mark them.

And the result — and this is why I am writing you — was for me an unparalleled experience, inexplicable in ordinary terms. I myself cannot fully account for it, except to be grateful that it happened to me. Some of the very qualities which I experienced in your course I found in this which I gave. I found myself liking these particular students as I have never liked any other group of persons, and I found — and they expressed this in their final report — that they themselves began to feel warm and kindly and accepting of one another. Orally and in their papers, they told of how moved they were, how much they learned, how well they felt. For me this was a brand new experience, and I was overwhelmed and humbled by it. I have had students who, I believe, respected and admired me, but I never had a classroom experience from which came such warmth and closeness. Incidentally, following your example, I avoided setting any fixed requirements in terms of reading or classroom preparation.

That the foregoing was not "biased perception" was evidenced from reports I got outside the classroom. The students had said such nice things about me that faculty members wanted to sit in the class. Best of all, the students at the end of the course wrote Dean

Benjamin Fine a letter in which they said the nicest things about me. And the Dean in turn wrote me to the same effect.

To say that I am overwhelmed by what happened only faintly reflects my feelings. I have taught for many years but I have never experienced anything remotely resembling what occurred. I, for my part, never have found in the classroom so much of the whole person coming forth, so deeply involved, so deeply stirred. Further, I question if in the traditional set-up, with its emphasis on subject matter, examinations, grades, there is, or there can be a place for the "becoming" person, with his deep and manifold needs, as he struggles to fulfill himself. But this is going far afield. I can only report to you what happened and to say that I am grateful and that I am also humbled by the experience. I would like you to know this, for again you have added to and enriched my life and being.*

* That this was not an isolated experience for Dr. Tenenbaum is indicated by a quotation from still another personal communication, many months later. He says: "With another group I taught, following the first one, similar attitudes developed, only they were more accentuated, because, I believe, I was more comfortable with the technique and, I hope, more expert. In this second group there was the same release of the person, the same exhilaration and excitement, the same warmth, the same mystery that attaches to a person as he succeeds in shedding portions of his skin. Students from my group told me that while attending other classes, their eyes would meet, drawn to one another, as if they were unique and apart, as if they were bound together by a special experience. In this second group, also, I found that the students had developed a personal closeness, so that at the end of the semester they talked of having annual reunions. They said that somehow or other they wanted to keep this experience alive and not lose one another. They also spoke of radical and fundamental changes in their person — in outlook, in values, in feelings, in attitudes both toward themselves and toward others."

16

The Implications of Client-Centered Therapy for Family Life

When I was asked, several years ago, to speak to a local group on any topic I wished, I decided to take a specific look at the changes in behavior exhibited by our clients in their family relationships. This paper was the result.

As an increasing number of our therapists and counselors have dealt with troubled individuals and groups, there has been agreement that our experience is relevant to, and has implications for, every area of interpersonal relationships. An attempt has been made to spell out some of the implications in certain areas — in the field of education, for example, in the area of group leadership, in the area of inter-group relationships — but we have never tried to make explicit what it means in family life. This is the realm with which I should like to deal now, trying to give as clear a picture as I can of what meanings a client-centered point of view seems to have for that closest of all interpersonal circles — the family group.

I do not wish to approach this from an abstract or theoretical level. What I wish to do is to present something of the changes our clients have experienced in their family relationships as they endeavor to work toward a more satisfactory life in their contacts with a therapist. I shall draw heavily on the verbatim statements of these people in order that you may get the flavor of their actual experience, and draw your own conclusions for yourself.

Although some of the experience of our clients seems to run counter to current concepts of what is involved in constructive family living, I am not particularly interested in arguing these differences. Also I am not particularly interested in setting up some model for family life in general, or in proposing the manner in which you should live in your family situation. I simply wish to present the gist of the experience of some very real people in some very real and often difficult family situations. Perhaps their struggles to live in a satisfying fashion will have some meaning for you.

What then, are some of the ways in which clients change in their family living, as a consequence of client-centered therapy?

More Expressive of Feeling

In the first place it is our experience that our clients gradually come to express more fully, to members of their families as well as to others, their true feelings. This applies to feelings that might be thought of as negative — resentment, anger, shame, jealousy, dislike, annoyance — as well as feelings which might be thought of as positive — tenderness, admiration, liking, love. It is as though the client discovers in therapy that it is possible to drop the mask he has been wearing, and become more genuinely himself. A husband finds himself becoming furiously angry with his wife, and expressing this anger, where before he had maintained — or thought he had maintained — a calm and objective attitude toward her behavior. It is as though the map of expression of feelings has come to match more closely the territory of the actual emotional experience. Parents and children, husbands and wives, come closer to expressing the feelings which really exist in them, rather than hiding their true feelings from the other person, or from the other person and themselves.

Perhaps an illustration or two would make this point more clear. A young wife, Mrs. M., comes for counseling. Her complaint is that her husband, Bill, is very formal and reserved with her, that he doesn't talk to her or share his thinking with her, is inconsiderate, that they are sexually incompatible and rapidly growing apart. As she talks out her attitudes the picture changes rather drastically. She expresses the deep guilt feeling which she has regarding her life before her marriage, when she had affairs with a number of men, mostly married men. She realizes that though with most people she is a gay and spontaneous person, with her husband she is stiff, controlled, lacking in spontaneity. She also sees herself as demanding that he be exactly what she wishes him to be. At this point counseling is interrupted by the counselor's absence from the city. She continues to write to the counselor expressing her feelings, and adding, "If I could only say these things to him (her husband) I could be myself at home. But what would that do to his trust in people? Would you find me repulsive if you were my husband and learned the truth? I wish I were a 'nice gal' instead of a 'Babe.' I've made such a mess of things."

This is followed by a letter from which a lengthy quotation seems justified. She tells how irritable she has been — how disagreeable she was when company dropped in one evening. After they left "I felt like a louse for behaving so badly. . . . I was still feeling sullen, guilty, angry at myself and Bill — and just about as blue as they come.

"So, I decided to do what I've been really wanting to do and putting off because I felt it was more than I could expect from any man — to tell Bill just what was making me act that terrible way. It was even harder than telling you — and that was hard enough. I couldn't tell it in such minute detail but I did manage to get out some of those sordid feelings about my parents and then even more about those 'damn' men. The nicest thing I've ever heard him say was 'Well, maybe I can help you there' — when speaking of my parents. And he was very accepting of the things I had done. I told him how I felt so inadequate in so many situations — because

I have never been allowed to do so many things — even to know how to play cards. We *talked, discussed,* and really got down deep into so many of both our feelings. I didn't tell him as completely about the men — their names, but I did give him an idea of about how many. Well, he was so understanding and things have cleared up so much that I TRUST HIM. I'm not afraid now to tell him those silly little illogical feelings that keep popping into my head. And if I'm not afraid then maybe soon those silly things will stop popping. The other evening when I wrote to you I was almost ready to pull out — I even thought of just leaving town. (Escaping the whole affair.) But I realized that I'd just keep running from it and not be happy until it was faced. We talked over children and though we've decided to wait until Bill is closer to finishing school, I'm happy with this arrangement. Bill feels as I do about the things we want to do for our children — and most important the things we *don't* want to do to them. So if you don't get any more desperate sounding letters, you know things are going along as okay as can be expected.

"Now, I'm wondering — have you known all along that that was the only thing I could do to bring Bill and me closer? That was the one thing I kept telling myself wouldn't be fair to Bill. I thought it would ruin his faith in me and in everyone. I had a barrier so big between Bill and me that I felt he was almost a stranger. The only way I pushed myself to do it was to realize that if I didn't at least try his response to the things that were bothering me, it wouldn't be fair to him — to leave him without giving him a chance to prove that he could be trusted. He proved even more than that to me — that he's been down in hell too with his feelings — about his parents, and a good many people in general."

I believe this letter needs no comment. It simply means to me that as she had experienced in therapy the satisfaction of being herself, of voicing her deep feelings, it became impossible for her to behave differently with her husband. She found that she had to be and express her own deepest feelings, even if this seemed to risk her marriage.

Another element in the experience of our clients is a somewhat subtle one. They find that, as in this instance, expression of feelings is a deeply satisfying thing, where formerly it has nearly always seemed destructive and disastrous. The difference seems to be due to this fact. When a person is living behind a front, a façade, his unexpressed feelings pile up to some explosion point, and are then apt to be triggered off by some specific incident. But the feelings which sweep over the person and are expressed at such a time — in a temper storm, in a deep depression, in a flood of self-pity, and the like — often have an unfortunate effect on all concerned because they are so inappropriate to the specific situation and hence seem so unreasonable. The angry flare-up over one annoyance in the relationship may actually be the pent-up or denied feelings resulting from dozens of such situations. But in the context in which it is expressed it is unreasonable and hence not understood.

Here is where therapy helps to break a vicious circle. As the client is able to pour out, in all their accumulated anguish, fury, or despair, the emotions which he has been feeling, and as he accepts these feelings as his own, they lose their explosiveness. Hence he is more able to express, in any specific family relationship, the feelings aroused by that relationship. Since they do not carry such an overload from the past, they are more appropriate, and more likely to be understood. Gradually the individual finds himself expressing his feelings when they occur, not at some much later point after they have burned and festered in him.

Relationships Can Be Lived on a Real Basis

There is another effect which counseling seems to have on the way our clients experience their family relationships. The client discovers, often to his great surprise, that a relationship can be lived on the basis of the real feelings, rather than on the basis of a defensive pretense. There is a deep and comforting significance to this, as we have already seen in the case of Mrs. M. To discover that feelings of shame and anger and annoyance can be expressed, and that the relationship still survives, is reassuring. To find that one can express tenderness and sensitivity and fearfulness and yet not be betrayed — this is a deeply strengthening thing. It seems that part of the reason this works out constructively is that in therapy the individual learns to recognize and express his feelings *as* his own feelings, not as a fact about another person. Thus, to say to one's spouse "What you are doing is all wrong," is likely to lead only to debate. But to say "I feel very much annoyed by what you're doing," is to state one fact about the speaker's feelings, a fact which no one can deny. It no longer is an accusation about another, but a feeling which exists in oneself. "You are to blame for my feelings of inadequacy" is a debatable point, but "I feel inadequate when you do thus and so" simply contributes a real fact about the relationship.

But it is not only at the verbal level that this operates. The person who accepts his own feelings within himself, finds that a relationship can be lived on the basis of these real feelings. Let me illustrate this with a series of excerpts from the recorded interviews with Mrs. S.

Mrs. S. lived with her ten year old daughter and her seventy year old mother, who dominated the household by her "poor health." Mrs. S. was controlled by her mother, and unable to control her daughter, Carol. She felt resentful of her mother, but could not express this, because "I have felt guilty all my life. I grew up feeling guilty because everything that I did I felt was a . . . in some way affecting my mother's health. . . . In fact, a few years ago, it came to the point where I was having dreams at night about . . . shaking my mother and . . . I'd . . . I got the feeling that I just wanted to push her out of the way. And . . . I can understand how Carol might feel. She doesn't dare . . . and neither do I."

Mrs. S. knows that most people think she would be much better off if she left her mother, but she cannot. "I know that if I do leave her, that I couldn't possibly be happy, I'd be so worried about her. And I'd feel so badly about leaving a poor old lady alone."

As she complains about the extent to which she is dominated and controlled, she begins to see the part she is playing, a cowardly part. "I feel that my hands are tied. Perhaps I'm at fault . . . more than mother is. In fact I know I am, but I've sort of become a coward where mother's concerned. I'll do anything to avoid one of the scenes that she puts on about little things."

As she understands herself better she comes to an inward conclusion to try to live in the relationship according to what she believes is right, rather than in terms of her mother's wishes. She reports this at the beginning of an interview. "Well, I've made a stupendous discovery, that perhaps it's been my fault entirely in overcompensating to mother . . . in other words, spoiling her. So I made up my mind like I do every morning, but I think this time it's gonna work, that I would try to . . . oh, to be calm and quiet, and . . . if she does go into one of her spells, to just more or less ignore it as you would a child who throws a tantrum just to get attention. So I tried it. And she got angry over some little thing. And she jumped up from the table and went into her room. Well, I didn't rush in and say, oh, I'm sorry, and beg her to come back, and I simply just ignored it. So in a few minutes, why, she came back and sat down and was a little sulky but she was over it. So I'm going to try that for a while and. . . ."

Mrs. S. realizes clearly that the basis for her new behavior is that she has come genuinely to accept her own feelings toward her mother. She says, "Well, why not face it? You see, I've been feeling so horrible, and thinking what a horrible person I was to resent my mother. Well, let's just say, okay, I resent her; and I'm sorry; but let's face it and I'll try to make the best of it."

As she accepts herself more she becomes much more able to meet some of her own needs as well as those of her mother. "There's a lot of things that I've wanted to do for years and that I'm just going to start to do. Now, mother can be alone till ten o'clock at night there. She has a telephone by her bed and . . . if a fire starts or something, there are neighbors, or if she becomes ill . . . so I'm going to take some night courses through the public schools you know, and I'm going to do a lot of things that I've wanted to do all my life, and have sort of been a martyr in staying home resenting it . . . that I had to, and thinking, oh, well, and not doing it. Well, I'm going to now. And I think after the first time I go, why, she'll be all right."

Her new found feelings are soon put to a test in the relationship with her mother. "My mother had a very severe heart attack the other day and I said, well, you'd better go to the hospital and . . . and you certainly need hospitalization; and I whipped her down to the doctor, and the doctor said her heart was fine and she oughta get out and have a little fun. So she's going to visit a friend for a week and see the shows and have a good time. So . . . actually when it came down to getting ready to go to the hospital, how cruel I am to her by contradicting her in front of Carol and all that sort of thing, why, then she backed down and when she was faced with the fact that she . . . and her heart's just as strong as a bull's, why, she thought she might as well use it to have some fun with. So that's fine. Working out fine."

Up to this point it might seem as though the relationship had improved for Mrs. S., but not for her mother. There is, however, another side to the picture. Somewhat later Mrs. S. says "I still am very, very sorry for mother. I would hate to be like she is. And another thing, you know, I just got to the point where I just hated mother; I couldn't stand to touch her, or . . . I mean . . . brush against her or something. I don't mean, just for the moment, while I was angry or anything. But . . . I've also found myself, oh, feeling a little affectionate toward her; two or three times I've gone in without even thinking, kissed her goodnight, and I used to just holler from the door. And . . . I've been feeling kindlier toward her; that resentment that I've had is going, along with the hold that she had over me, you see. So . . . that, I noticed that yesterday when I was helping her get ready and so forth; I fixed her hair and there was the longest time I couldn't stand to touch her; and I was doing her hair in pin curls and so forth; and I . . . it suddenly came to me, well, now this doesn't bother me a bit; in fact it's kind of fun."

These excerpts seem to me to portray a pattern of change in family relationships which is very familiar to us. Mrs. S. feels, though she hardly dares admit it even to herself, resentful of her mother and as though she had no rights of her own. It seems as though nothing but difficulty could result from letting these feelings exist openly in the relationship. Yet as she tentatively permits them to enter the situation she finds herself acting with more assurance, more integrity. The relationship improves rather than deteriorates. Most surprising of all, when the relationship is lived on the basis of the real feelings, she finds that resentment and hate are not the only feelings she has toward her mother. Fondness, affection and enjoyment are also feelings which enter the relationship. It seems clear that there may be moments of discord, dislike, and anger between the two. But there will also be respect and understanding and liking. They seem to have learned what many other clients have also learned, that a relationship does not have to be lived on a basis of pretense, but can be lived on the basis of the fluctuating variety of feelings which actually exist.

It may seem, from the illustrations I have chosen, that it is only negative feelings which are difficult to express or live. This is far from true. Mr. K., a young professional man, found it fully as difficult to discover the positive feelings which lay beneath his façade, as the negative. A brief excerpt will indicate the changed quality of his relationship with his three-year-old daughter.

He says, "The thing that I was thinking about as I rode down here was — how differently I see our little girl — I was playing with her this morning — and — we just, ah, well — why is it so hard for me to get words out now? This was a really wonderful experience — very warm, and it was a happy and pleasant thing, and it seems that I saw and felt her so close to me. Here's what I think is significant — before, I could talk about Judy. I could say positive things about her and funny little things she'd do and just talk about her as though I were and felt like a real happy father, but there was some unreal quality . . . as though I was just saying these things because I *should* be feeling this stuff and this is the way a father *should* talk about his daughter but somehow this wasn't really true because I did have these negative and mixed up feelings about her. Now I do think she is the most wonderful kid in the world."

T: "Before, you felt as though 'I *should* be a happy father' — this morning you *are* a happy father. . . ."

"It certainly felt that way this morning. She just rolled around on the bed . . . and then she asked me if I wanted to go to sleep again and I said okay and then she said well, I'll go get my blankets . . . and then she told me a story . . . about three stories in one . . . all jumbled up and . . . it just felt like *this* is what I *really* want . . . I *want* to have this experience. It felt that I was . . . I felt grown up, I guess. I felt that I was a man . . . now this sounds strange, but it did feel as though I was a grownup responsible loving father, who was big enough, and serious enough, and also happy enough to be the father of this child. Whereas before I did feel weak and maybe almost undeserving, ineligible to be that important, because it is a very important thing to be a father."

He has found it possible to accept positive feelings toward himself as a good father, and to fully accept this warm love for his little girl. He no longer has to pretend he loves her, fearful that some different feeling may be lurking underneath.

I think it will not surprise you that shortly after this he told how he could be much more free in expressing anger and annoyance at his little daughter, also. He is learning that the feelings which exist are good enough to live by. They do not have to be coated with a veneer.

Improvement in Two-Way Communication

Experience in therapy seems to bring about another change in the way our clients live in their family relationships. They learn something about how to initiate and maintain real two-way communication. To understand another person's thoughts and feelings thoroughly, with the meanings they have for him, and to be thoroughly understood by this other person in return — this is one of the most rewarding of human experiences, and all too rare. Individuals who have come to us for therapy often report their pleasure in discovering that such genuine communication is possible with members of their own families.

In part this seems to be due, quite directly, to their experience of communication with the counselor. It is such a relief, such a blessed relaxation of defenses, to find oneself understood, that the individual wishes to create this atmosphere for others. To find, in the therapeutic relationship that one's most awful thoughts, one's most bizarre and abnormal feelings, one's most ridiculous dreams and hopes, one's most evil behaviors, can all be understood by another, is a tremendously releasing experience. One begins to see it as a resource he could extend to others.

But there appears to be an even more fundamental reason why these clients can understand members of their families. When we are living behind a façade, when we are trying to act in ways that are not in accord with our feelings, then we dare not listen freely to another. We must always have our guard up, lest he pierce the pretense of our façade. But when a client is living in the way I have been describing, when he tends to express his real feelings in the situation in which they occur, when his family relationships are lived on the basis of the feelings which actually exist, then he is no longer defensive and he can really listen to, and understand, another member of his family. He can let himself see how life appears to this other person.

Something of what I am saying may be illustrated from the experience of Mrs. S., the woman quoted in the preceding section. In a followup contact after the conclusion of her interviews, Mrs. S. was asked to give some of her own reactions to her experience. She says, "I didn't feel at first that it was counseling. You know? I thought, well, I'm just talking, but . . . by giving it a little thought, I realize that it is counseling and of the very best kind, because I've had advice, and excellent advice from doctors and family and friends and . . . it's never worked. And I think in order to reach people, you can't put up barriers and things of that sort, because then you don't get the true reaction. . . . But I've given it a great deal of thought and I'm sort of working it with Carol a little bit now (laughing) or trying to, you know. And . . . grandma says to her, how can you be so mean to your poor sick old grandmother, you know. And I just know how Carol feels. She just wants to hit her because she's so terrible! But I sort of haven't been saying too much to Carol or trying to guide her. But I've been trying to draw her out . . . let her feel that I'm with her and behind her, no matter what she does. And let her tell me how she feels, and her little reactions to things, and it's working out fine. She has told me, oh, grandma's been old and sick for so long, mother. And I said, yes. And I don't condemn her nor do I praise her, and so she is, just in this short time beginning to . . . oh, get little things off her mind and . . . without my probing or trying to . . . so it's sort of working on her. And it seems to be working on mother a little bit too."

I think we may say of Mrs. S. that having accepted her own feelings, and having been more willing to express them and to live in them, she now finds more willingness on her own part to understand her daughter and her mother, and to feel empathically their own reactions to life. She is sufficiently free of defensiveness to be able to listen in an accepting manner, and to sense the way life feels to them. This kind of development seems characteristic of the change which occurs in the family life of our clients.

Willingness for Another to be Separate

There is one final tendency which we have noticed and which I would like to describe. It is quite noticeable that our clients tend in the direction of permitting each member of the family to have his own feelings and to be a separate person. This may seem a strange statement, but it is actually a most radical step. Many of us are perhaps unaware of the tremendous pressure we tend to put on our wives, our husbands, our children, to have the same feelings we do. It is often as though we said, "If you want me to love you, then you must have the same feelings I do. If I feel your behavior is bad, you must feel so too. If I feel a certain goal is desirable, you must feel so too." Now the tendency which we see in our clients is the opposite of this. There is a willingness for the other person to have different feelings, different values, different goals. In short, there is a willingness for him to be a separate person.

It is my belief that this tendency develops as the person discovers that he can trust his own feelings and reactions — that his own deep impulses are not destructive or catastrophic, and that he himself need not be guarded, but can meet life on a real basis. As he thus learns that he can trust himself, with his own uniqueness, he becomes more able to trust his wife, or his child, and to accept the unique feelings and values which exist in this other person.

Something of what I mean is contained in letters from a woman and her husband. They are friends of mine and had obtained a copy of a book I had written because they were interested in what I was doing. But the effect of the book seemed to be similar to therapy. The wife wrote me and included in her letter a paragraph giving her reactions. "Lest you think that we are completely frivolous, we have been reading *Client-Centered Therapy*. I have almost finished it. Most of the usual things you say about books don't apply, at least for me. In fact it was pretty close to a counseling experience. It set me to thinking about some of the unsatisfactory relationships in our family, particularly my attitude toward Phillip (her 14-year-old son). I realized that I hadn't shown him any real love for a long time, because I was so resentful of his apparent indifference in trying to measure up to any of the standards that I have always thought were important. Since I have stopped taking most of the responsibility for his goals, and have responded to him as a person, as I always have to Nancy, for instance, it is surprising what changes have appeared in his attitudes. Not earth-shaking — but a heartwarming beginning. We no longer heckle him about his school work, and the other day he volunteered that he had gotten an S — satisfactory grade — on a math exam. The first time this year."

A few months later I heard from her husband. "You wouldn't recognize Phil. . . . While he is hardly garrulous, he is not nearly the sphinx that he was, and he is doing much better in school, although we do not expect him to be graduated cum laude. You should take a great deal of credit for his improvement, because he began to blossom when I finally began to trust him to be himself, and ceased trying to mold him into the glorified image of his father at a similar age. Oh to undo our past errors!"

This concept of trusting the individual to be himself has come to have a great deal of meaning to me. I sometimes fantasy about what it would mean if a child were treated in this fashion from the first. Suppose a child were permitted to have his own unique feelings — suppose he never had to disown his feelings in order to be loved. Suppose his parents were free to have and express their own unique feelings, which often would be different from his, and often different between themselves. I like to think of all the meanings that such an experience would have. It would mean that the child would grow up respecting himself as a unique person. It would mean that even when his behavior had to be thwarted, he could retain open "ownership" of his feelings. It would mean that his behavior would be a realistic balance, taking into account his own feelings and the known and open feelings of others. He would, I believe, be a responsible and self-directing individual, who would never need to conceal his feelings from himself, who would never need to live behind a façade. He would be relatively free of the maladjustments which cripple so many of us.

The General Picture

If I have been able correctly to discern the trends in the experience of our clients, then client-centered therapy seems to have a number of implications for family life. Let me attempt to restate these in somewhat more general form.

It appears that an individual finds it satisfying in the long run to express any strong or persistent emotional attitudes in the situation in which they arise, to the person with whom they are concerned, and to the depth to which they exist. This is more satisfying than refusing to admit that such feelings exist, or permitting them to pile up to an explosive degree, or directing them toward some situation other than the one in which they arose.

It seems that the individual discovers that it is more satisfying in the long run to live a given family relationship on the basis of the real interpersonal feelings which exist, rather than living the relationship on the basis of a pretense. A part of this discovery is that the fear that the relationship will be destroyed if the true feelings are admitted, is usually unfounded, particularly when the feelings are expressed as belonging to oneself, not as stating something about the other person.

Our clients find that as they express themselves more freely, as the surface character of the relationship matches more closely the fluctuating attitudes which underlie it, they can lay aside some of their defenses and truly listen to the other person. Often for the first time they begin to understand how the other person feels, and why he feels that way. Thus mutual understanding begins to pervade the interpersonal interaction.

Finally, there is an increasing willingness for the other person to be himself. As I am more willing to be myself, I find I am more ready to permit you to be yourself, with all that that implies. This means that the family circle tends in the direction of becoming a number of separate and unique persons with individual goals and values, but bound together by the real feelings — positive and negative — which exist between them, and by the satisfying bond of mutual understanding of at least a portion of each other's private worlds.

It is in these ways, I believe, that a therapy which results in the individual becoming more fully and more deeply himself, results also in his finding greater satisfaction in realistic family relationships which likewise promote the same end — that of facilitating each member of the family in the process of discovering, and becoming, himself.

17

Dealing With Breakdowns in Communication— Interpersonal and Intergroup

In point of time, this paper is the earliest in the book. It was written in 1951 for presentation at the Centennial Conference on Communications at Northwestern University, where it was given the title, "Communication: Its Blocking and Its Facilitation." It has since been reprinted a half-dozen times, by different groups and in different journals, including the Harvard Business Review and ETC, the journal of the Society for General Semantics.

Although some of its illustrations now appear a bit dated, I am including it because it makes what I feel is an important point regarding group tensions, national and international. The suggestion regarding Russian–U.S. tensions appeared hopelessly idealistic at that time. Now I believe it would be accepted by many as good sense.

IT MAY SEEM CURIOUS that a person whose whole professional effort is devoted to psychotherapy should be interested in problems of communication. What relationship is there between providing therapeutic help to individuals with emotional maladjustments and the concern of this conference with obstacles to communication? Actually the relationship is very close indeed. The whole task of psychotherapy is the task of dealing with a failure in communication. The emotionally maladjusted person, the "neurotic," is in difficulty first, because communication within himself has broken down, and second because, as a result of this, his communication with others has been damaged. If this sounds somewhat strange to you, then let me put it in other terms. In the "neurotic" individual, parts of himself which have been termed unconscious, or repressed, or denied to awareness, become blocked off so that they no longer communicate themselves to the conscious or managing part of himself. As long as this is true, there are distortions in the way he communicates himself to others, and so he suffers both within himself, and in his interpersonal relations. The task of psychotherapy is to help the person achieve, through a special relationship with a therapist, good communication within himself. Once this is achieved he can communicate more freely and more effectively with others. We may say then that psychotherapy is good communication, within and between men. We may also turn that statement around and it will still be true. Good communication, free communication, within or between men, is always therapeutic.

It is, then, from a background of experience with communication in counseling and psychotherapy, that I want to present to you tonight two ideas. I wish to state what I believe is one of the major factors in blocking or impeding communication, and then I wish to present what in our experience has proven to be a very important way of improving or facilitating communication.

I would like to propose, as an hypothesis for consideration, that the major barrier to mutual interpersonal communication is our very natural tendency to judge, to evaluate, to approve or disapprove, the statement of the other person, or the other group. Let me illustrate my meaning with some very simple examples. As you leave the meeting tonight, one of the statements you are likely to hear is, "I didn't like that man's talk." Now what do you respond? Almost invariably your reply will be either approval or disapproval of the attitude expressed. Either you respond, "I didn't either. I thought it was terrible," or else you tend to reply, "Oh, I thought it was really good." In other words, your primary reaction is to evaluate what has just been said to you, to evaluate it from *your* point of view, your own frame of reference.

Or take another example. Suppose I say with some feeling, "I think the Republicans are behaving in ways that show a lot of good sound sense these days," what is the response that arises in your mind as you listen? The overwhelming likelihood is that it will be evaluative. You will find yourself agreeing, or disagreeing, or making some judgment about me such as "He must be a conservative," or "He seems solid in his thinking." Or let us take an illustration from the international scene. Russia says vehemently, "The treaty with Japan is a war plot on the part of the United States." We rise as one person to say "That's a lie!"

This last illustration brings in another element connected with my hypothesis. Although the tendency to make evaluations is common in almost all interchange of language, it is very much heightened in those situations where feelings and emotions are deeply involved. So the stronger our feelings the more likely it is that there will be no mutual element in the communication. There will be just two ideas, two feelings, two judgments, missing each other in psychological space. I'm sure you recognize this from your own experience. When you have not been emotionally involved yourself, and have listened to a heated discussion, you often go away thinking, "Well, they actually weren't talking about the same thing." And they were not. Each was making a judgment, an evaluation, from his own frame of reference. There was really nothing which could be called communication in any genuine sense. This tendency to react to any emotionally meaningful statement by forming an evaluation of it from our own point of view, is, I repeat, the major barrier to interpersonal communication.

But is there any way of solving this problem, of avoiding this barrier? I feel that we are making exciting progress toward this goal and I would like to present it as simply as I can. Real communication occurs, and this evaluative tendency is avoided, when we listen with understanding. What does this mean? It means to see the expressed idea and attitude from the other person's point of view, to sense how it feels to him, to achieve his frame of reference in regard to the thing he is talking about.

Stated so briefly, this may sound absurdly simple, but it is not. It is an approach which we have found extremely potent in the field of psychotherapy. It is the most effective agent we know for altering the basic personality structure of an individual, and improving his relationships and his communications with others. If I can listen to what he can tell me, if I can understand how it seems to him, if I can see its personal meaning for him, if I can sense the emotional flavor which it has for him, then I will be releasing potent forces of change in him. If I can really understand how he hates his father, or hates the university, or hates communists — if I can catch the flavor of his fear of insanity, or his fear of atom bombs, or of Russia — it will be of the greatest help to him in altering those very hatreds and fears, and in establishing realistic and harmonious relationships with the very people and situations toward which he has felt hatred and fear. We know from our research that such empathic understanding — understanding *with* a person, not *about* him — is such an effective approach that it can bring about major changes in personality.

Some of you may be feeling that you listen well to people, and that you have never seen such results. The chances are very great indeed that your listening has not been of the type I have described. Fortunately I can suggest a little laboratory experiment which you can try to test the quality of your understanding. The next time you get into an argument with your wife, or your friend, or with a small group of friends, just stop the discussion for a moment and for an experiment, institute this rule. "Each person can speak up for himself only *after* he has first restated the ideas and feelings of the previous speaker accurately, and to that speaker's satisfaction." You see what this would mean. It would simply mean that before presenting your own point of view, it would be necessary for you to really achieve the other speaker's frame of reference — to understand his thoughts and feelings so well that you could summarize them for him. Sounds simple, doesn't it? But if you try it you will discover it is one of the most difficult things you have ever tried to do. However, once you have been able to see the other's point of view, your own comments will have to be drastically revised. You will also find the emotion going out of the discussion, the differences being reduced, and those differences which remain being of a rational and understandable sort.

Can you imagine what this kind of an approach would mean if it were projected into larger areas? What would happen to a labor-management dispute if it was conducted in such a way that labor, without necessarily agreeing, could accurately state management's point of view in a way that management could accept; and management, without approving labor's stand, could state labor's case in a way that labor agreed was accurate? It would mean that real communication was established, and one could practically guarantee that some reasonable solution would be reached.

If then this way of approach is an effective avenue to good communication and good relationships, as I am quite sure you will agree if you try the experiment I have mentioned, why is it not more widely tried and used? I will try to list the difficulties which keep it from being utilized.

In the first place it takes courage, a quality which is not too widespread. I am indebted to Dr. S. I. Hayakawa, the semanticist, for pointing out that to carry on psychotherapy in this fashion is to take a very real risk, and that courage is required. If you really understand another person in this way, if you are willing to enter his private world and see the way life appears to him, without any attempt to make evaluative judgments, you run the risk of being changed yourself. You might see it his way, you might find yourself influenced in your attitudes or your personality. This risk of being changed is one of the most frightening prospects most of us can face. If I enter, as fully as I am able, into the private world of a neurotic or psychotic individual, isn't there a risk that I might become lost in that world? Most of us are afraid to take that risk. Or if we had a Russian communist speaker here tonight, or Senator Joseph McCarthy, how many of us would dare to try to see the world from each of these points of view? The great majority of us could not *listen;* we would find ourselves compelled to *evaluate*, because listening would seem too dangerous. So the first requirement is courage, and we do not always have it.

But there is a second obstacle. It is just when emotions are strongest that it is most difficult to achieve the frame of reference of the other person or group. Yet this is the time the attitude is most needed, if communication is to be established. We have not found this to be an insuperable obstacle in our experience in psychotherapy. A third party, who is able to lay aside his own feelings and evaluations, can assist greatly by listening with understanding to each person or group and clarifying the views and attitudes each holds. We have found this very effective in small groups in which contradictory or antagonistic attitudes exist. When the parties to a dispute realize that they are being understood, that someone sees how the situation seems to them, the statements grow less exaggerated and less defensive, and it is no longer necessary to maintain the attitude, "I am 100 per cent right and you are 100 per cent wrong." The influence of such an understanding catalyst in the group permits the members to come closer and closer to the objective truth involved in the relationship. In this way mutual communication is established and some type of agreement becomes much more possible. So we may say that though heightened emotions make it much more difficult to understand *with* an opponent, our experience makes it clear that a neutral, understanding, catalyst type of leader or therapist can overcome this obstacle in a small group.

This last phrase, however, suggests another obstacle to utilizing the approach I have described. Thus far all our experience has been with small face-to-face groups — groups exhibiting industrial tensions, religious tensions, racial tensions, and therapy groups in which many personal tensions are present. In these small groups our experience, confirmed by a limited amount of research, shows that a listening, empathic approach leads to improved communication, to greater acceptance of others and by others, and to attitudes which are more positive and more problem-solving in nature. There is a decrease in defensiveness, in exaggerated statements, in evaluative and critical behavior. But these findings are from small groups. What about trying to achieve understanding between larger groups that are geographically remote? Or between face-to-face groups who are not speaking for themselves, but simply as representatives of others, like the delegates at the United Nations? Frankly we do not know the answers to these questions. I believe the situation might be put this way. As social scientists we have a tentative test-tube solution of the problem of breakdown in communication. But to confirm the validity of this test-tube solution, and to adapt it to the enormous problems of communication breakdown between classes, groups, and nations, would involve additional funds, much more research, and creative thinking of a high order.

Even with our present limited knowledge we can see some steps which might be taken, even in large groups, to increase the amount of listening *with*, and to decrease the amount of evaluation *about*. To be imaginative for a moment, let us suppose that a therapeutically oriented international group went to the Russian leaders and said, "We want to achieve a genuine understanding of your views and even more important, of your attitudes and feelings, toward the United States. We will summarize and resummarize these views and feelings if necessary, until you agree that our description represents the situation as it seems to you." Then suppose they did the same thing with the leaders in our own country. If they then gave the widest possible distribution to these two views, with the feelings clearly described but not expressed in name-calling, might not the effect be very great? It would not guarantee the type of understanding I have been describing, but it would make it much more possible. We can understand the feelings of a person who hates us much more readily when his attitudes are accurately described to us by a neutral third party, than we can when he is shaking his fist at us.

But even to describe such a first step is to suggest another obstacle to this approach of understanding. Our civilization does not yet have enough faith in the social sciences to utilize their findings. The opposite is true of the physical sciences. During the war when a test-tube solution was found to the problem of synthetic rubber, millions of dollars and an army of talent was turned loose on the problem of using that finding. If synthetic rubber could be made in milligrams, it could and would be made in the thousands of tons. And it was. But in the social science realm, if a way is found of facilitating communication and mutual understanding in small groups, there is no guarantee that the finding will be utilized. It may be a generation or more before the money and the brains will be turned loose to exploit that finding.

In closing, I would like to summarize this small-scale solution to the problem of barriers in communication, and to point out certain of its characteristics.

I have said that our research and experience to date would make it appear that breakdowns in communication, and the evaluative tendency which is the major barrier to communication, can be avoided. The solution is provided by creating a situation in which each of the different parties comes to understand the other from the *other's* point of view. This has been achieved, in practice, even when feelings run high, by the influence of a person who is willing to understand each point of view empathically, and who thus acts as a catalyst to precipitate further understanding.

This procedure has important characteristics. It can be initiated by one party, without waiting for the other to be ready. It can even be initiated by a neutral third person, providing he can gain a minimum of cooperation from one of the parties.

This procedure can deal with the insincerities, the defensive exaggerations, the lies, the "false fronts" which characterize almost every failure in communication. These defensive distortions drop away with astonishing speed as people find that the only intent is to understand, not judge.

This approach leads steadily and rapidly toward the discovery of the truth, toward a realistic appraisal of the objective barriers to communication. The dropping of some defensiveness by one party leads to further dropping of defensiveness by the other party, and truth is thus approached.

This procedure gradually achieves mutual communication. Mutual communication tends to be pointed toward solving a problem rather than toward attacking a person or group. It leads to a situation in which I see how the problem appears to you, as well as to me, and you see how it appears to me, as well as to you. Thus accurately and realistically defined, the problem is almost certain to yield to intelligent attack, or if it is in part insoluble, it will be comfortably accepted as such.

This then appears to be a test-tube solution to the breakdown of communication as it occurs in small groups. Can we take this small scale answer, investigate it further, refine it, develop it and apply it to the tragic and well-nigh fatal failures of communication which threaten the very existence of our modern world? It seems to me that this is a possibility and a challenge which we should explore.

18

A Tentative Formulation of a General Law of Interpersonal Relationships

During a recent summer I gave some thought to a theoretical problem which had tantalized me: Would it be possible to formulate, in one hypothesis, the elements which make any relationship either growth-facilitating or the reverse. I worked out a short document for myself, and had occasion to try it out on a workshop group and some industrial executives with whom I was conferring. It seemed to be of interest to all, but most stimulating to the industrial leaders who discussed it pro and con in terms of such problems as: supervisor-supervisee relationships; labor-management relationships; executive training; interpersonal relations among top management.

I regard this as a highly tentative document, and am not at all sure of its adequacy. I include it because many who have read it have found it provocative, and because publication of it may inspire research studies which would begin to test its validity.

I HAVE MANY TIMES ASKED myself how our learnings in the field of psychotherapy apply to human relationships in general. During recent years I have thought much about this issue and attempted to state a theory of interpersonal relationships as a part of the larger structure of theory in client-centered therapy (1, Sec. IV). This present document undertakes to spell out, in a somewhat different way, one of the aspects of that theory. It endeavors to look at a perceived underlying orderliness in all human relationships, an order which determines whether the relationship will make for the growth, enhancement, openness, and development of both individuals or whether it will make for inhibition of psychological growth, for defensiveness and blockage in both parties.

The Concept of Congruence

Fundamental to much of what I wish to say is the term "congruence." This construct has been developed to cover a group of phenomena which seem important to therapy and to all interpersonal interaction. I would like to try to define it.

Congruence is the term we have used to indicate an accurate matching of experiencing and awareness. It may be still further extended to cover a matching of experience, awareness, and communication. Perhaps the simplest example is an infant. If he is experiencing hunger at the physiological and visceral level, then his awareness appears to match this experience, and his communication is also congruent with his experience. He is hungry and dissatisfied, and this is true of him at all levels. He is at this moment integrated or unified in being hungry. On the other hand if he is satiated and content this too is a unified congruence, similar at the visceral level, the level of awareness and the level of communication. He is one unified person all the way through, whether we tap his experience at the visceral level, the level of his awareness, or the level of communication. Probably one of the reasons why most people respond to infants is that they are so completely genuine, integrated or congruent. If an infant expresses affection or anger or contentment or fear there is no doubt in our minds that he *is* this experience, all the way through. He is transparently fearful or loving or hungry or whatever.

For an example of incongruence we must turn to someone beyond the stage of infancy. To pick an easily recognizable example take the man who becomes angrily involved in a group discussion. His face flushes, his tone communicates anger, he shakes his finger at his opponent. Yet when a friend says, "Well, let's not get angry about this," he replies, with evident sincerity and surprise, "I'm not angry! I don't have any *feeling* about this at all! I was just pointing out the logical facts." The other men in the group break out in laughter at this statement.

What is happening here? It seems clear that at a physiological level he is experiencing anger. This is not matched by his awareness. Consciously he is *not* experiencing anger, nor is he communicating this (so far as he is consciously aware). There is a real incongruence between experience and awareness, and between experience and communication.

Another point to be noted here is that his communication is actually ambiguous and unclear. In its words it is a setting forth of logic and fact. In its tone, and in the accompanying gestures, it is carrying a very different message — "I am angry at you." I believe this ambiguity or contradictoriness of communication is always present when a person who is at that moment incongruent endeavors to communicate.

Still another facet of the concept of incongruence is illustrated by this example. The individual himself is not a sound judge of his own degree of congruence. Thus the laughter of the group indicates a clear consensual judgment that the man is *experiencing* anger, whether or not he thinks so. Yet in his own awareness this is not true. In other words it appears that the degree of congruence cannot be evaluated by the person himself at that moment. We may make progress in learning to measure it from an external frame of reference. We have also learned much about incongruence from the person's own ability to recognize incongruence in himself in the past. Thus if the man of our example were in therapy, he might look back on this incident in the acceptant safety of the therapeutic hour and say, "I realize now I was terribly angry at him, even though at the time I thought I was not." He has, we say, come to recognize that his defensiveness at that moment kept him from being aware of his anger.

One more example will portray another aspect of incongruence. Mrs. Brown, who has been stifling yawns and looking at her watch for hours, says to her hostess on departing, "I enjoyed this evening *so* much. It was a delightful party." Here the incongruence is not between experience and awareness. Mrs. Brown is well aware that she is bored. The incongruence is between awareness and communication. Thus it might be noted that when there is an incongruence between experience and awareness, it is usually spoken of as defensiveness, or denial to awareness. When the incongruence is between awareness and communication it is usually thought of as falseness or deceit.

There is an important corollary of the construct of congruence which is not at all obvious. It may be stated in this way. If an individual is at this moment entirely congruent, his actual physiological experience being accurately represented in his awareness, and his communication being accurately congruent with his awareness, then his communication could never contain an expression of an external fact. If he was congruent he could not say, "That rock is hard"; "He is stupid"; "You are bad"; or "She is intelligent." The reason for this is that we never *experience* such "facts." Accurate awareness of *experience* would always be expressed as feelings, perceptions, meanings from an internal frame of reference. I never *know* that he is stupid or you are bad. I can only perceive that you seem this way to me. Likewise, strictly speaking I do not *know* that the rock is hard, even though I may be very sure that I *experience* it as hard if I fall down on it. (And even then I can permit the physicist to perceive it as a very permeable mass of high-speed atoms and molecules.) If the person is thoroughly congruent then it is clear that all of his communication would necessarily be put in a context of personal perception. This has very important implications.

As an aside it might be mentioned that for a person always to speak from a context of personal perception does not necessarily imply congruence, since any mode of expression *may* be used as a type of defensiveness. Thus the person in a moment of congruence would necessarily communicate his perceptions and feelings as being these, and not as being *facts* about another person or the outside world. The reverse does not necessarily hold, however.

Perhaps I have said enough to indicate that this concept of congruence is a somewhat complex concept with a number of characteristics and implications. It is not easily defined in operational terms, though some studies have been completed and others are in process which do provide crude operational indicators of what is being experienced, as distinct from the awareness of that experience. It is believed that further refinements are possible.

To conclude our definition of this construct in a much more commonsense way, I believe all of us tend to recognize congruence or incongruence in individuals with whom we deal. With some individuals we realize that in most areas this person not only consciously means exactly what he says, but that his deepest feelings also match what he is expressing, whether it is anger or competitiveness or affection or cooperativeness. We feel that "we know exactly where he stands." With another individual we recognize that what he is saying is almost certainly a front, a façade. We wonder what he *really* feels. We wonder if *he* knows what he feels. We tend to be wary and cautious with such an individual.

Obviously then different individuals differ in their degree of congruence, and the same individual differs at different moments in degree of congruence, depending on what he is experiencing and whether he can accept this experience in his awareness, or must defend himself against it.

Relating Congruence to Communication in Interpersonal Relationships

Perhaps the significance of this concept for interpersonal interaction can be recognized if we make a few statements about a hypothetical Smith and Jones.

1. Any communication of Smith to Jones is marked by some degree of congruence in Smith. This is obvious from the above.

2. The greater the congruence of experience, awareness, and communication in Smith, the more it is likely that Jones will experience it as a *clear* communication. I believe this has been adequately covered. If all the cues from speech, tone and gesture are unified because they spring from a congruence and unity in Smith, then there is much less likelihood that these cues will have an ambiguous or unclear meaning to Jones.

3. Consequently, the more clear the communication from Smith, the more Jones responds with clarity. This is simply saying that

even though Jones might be quite *in*congruent in his experiencing of the topic under discussion, nevertheless his response will have *more* clarity and congruence in it than if he had experienced Smith's communication as ambiguous.

4. The more that Smith is congruent in the topic about which they are communicating, the less he has to defend himself against in this area, and the more able he is to listen accurately to Jones' response. Putting it in other terms, Smith has expressed what he genuinely feels. He is therefore more free to listen. The less he is presenting a façade to be defended, the more he can listen accurately to what Jones is communicating.

5. But to this degree, then, Jones feels empathically understood. He feels that in so far as he has expressed himself, (and whether this is defensively or congruently) Smith has understood him pretty much as he sees himself, and as he perceives the topic under consideration.

6. For Jones to feel understood is for him to experience positive regard for Smith. To feel that one is understood is to feel that one has made some kind of a positive difference in the experience of another, in this case of Smith.

7. But to the degree that Jones (a) experiences Smith as congruent or integrated in this relationship; (b) experiences Smith as having positive regard for him; (c) experiences Smith as being empathically understanding; to that degree the conditions of a therapeutic relationship are established. I have tried in another paper (2) to describe the conditions which our experience has led us to believe are necessary and sufficient for therapy, and will not repeat that description here.

8. To the extent that Jones is experiencing these characteristics of a therapeutic relationship, he finds himself experiencing fewer barriers to communication. Hence he tends to communicate himself more as he is, more congruently. Little by little his defensiveness decreases.

9. Having communicated himself more freely, with less of defensiveness, Jones is now more able to listen accurately, without a need for defensive distortion, to Smith's further communication. This is a repetition of step 4, but now in terms of Jones.

10. To the degree that Jones is able to listen, Smith now feels

empathically understood (as in step 5 for Jones); experiences Jones' positive regard (a parallel to step 6); and finds himself experiencing the relationship as therapeutic (in a way parallel to step 7). Thus Smith and Jones have to some degree become reciprocally therapeutic for each other.

11. This means that to some degree the process of therapy occurs in each and that the outcomes of therapy will to that same degree occur in each; change in personality in the direction of greater unity and integration; less conflict and more energy utilizable for effective living; change in behavior in the direction of greater maturity.

12. The limiting element in this chain of events appears to be the introduction of threatening material. Thus if Jones in step 3 includes in his more congruent response new material which is outside of the realm of Smith's congruence, touching an area in which Smith is *in*congruent, then Smith may not be able to listen accurately, he defends himself against hearing what Jones is communicating, he responds with communication which is ambiguous, and the whole process described in these steps begins to occur in reverse.

A Tentative Statement of a General Law

Taking all of the above into account, it seems possible to state it far more parsimoniously as a generalized principle. Here is such an attempt.

Assuming (a) a minimal willingness on the part of two people to be in contact; (b) an ability and minimal willingness on the part of each to receive communication from the other; and (c) assuming the contact to continue over a period of time; then the following relationship is hypothesized to hold true.

> The greater the congruence of experience, awareness and communication on the part of one individual, the more the ensuing relationship will involve: a tendency toward reciprocal communication with a quality of increasing congruence; a tendency toward more mutually accurate understanding of the communications; improved psychological adjustment and functioning in both parties; mutual satisfaction in the relationship.

> Conversely the greater the communicated *incongruence* of experience and awareness, the more the ensuing relationship will involve: further communication with the same quality; disintegration of accurate understanding, less adequate psychological adjustment and functioning in both parties; and mutual dissatisfaction in the relationship.

With probably even greater formal accuracy this general law could be stated in a way which recognizes that it is the perception of the *receiver* of communication which is crucial. Thus the hypothesized law could be put in these terms, assuming the same pre-conditions as before as to willingness to be in contact, etc.

> The more that Y experiences the communication of X as a congruence of experience, awareness, and communication, the more the ensuing relationship will involve: (etc, as stated above.)

Stated in this way this "law" becomes an hypothesis which it should be possible to put to test, since Y's *perception* of X's communication should not be too difficult to measure.

The Existential Choice

Very tentatively indeed I would like to set forth one further aspect of this whole matter, an aspect which is frequently very real in the therapeutic relationship, and also in other relationships, though perhaps less sharply noted.

In the actual relationship both the client and the therapist are frequently faced with the existential choice, "Do I dare to communicate the full degree of congruence which I feel? Do I dare match my experience, and my awareness of that experience, with my communication? Do I dare to communicate myself as I am or must my communication be somewhat less than or different from this?" The sharpness of this issue lies in the often vividly foreseen possibility of threat or rejection. To communicate one's full awareness of the relevant experience is a risk in interpersonal relationships. It seems to me that it is the taking or not taking of this risk which determines whether a given relationship becomes more and more mutually therapeutic or whether it leads in a disintegrative direction.

To put it another way. I cannot choose whether my awareness will be congruent with my experience. This is answered by my need for defense, and of this I am not aware. But there is a continuing existential choice as to whether my communication will be congruent with the awareness I *do* have of what I am experiencing. In this moment-by-moment choice in a relationship may lie the answer as to whether the movement is in one direction or the other in terms of this hypothesized law.

REFERENCES

1. Rogers, Carl R. A theory of therapy, personality and interpersonal relationships. In Koch, S. (Ed.). *Psychology: A Study of a Science*, vol. III. New York: McGraw-Hill, 1959, 184–256.

2. Rogers, Carl R. The necessary and sufficient conditions of therapeutic personality change, *J. Consult. Psychol.*, *21*, 95–103.

19

Toward a Theory of Creativity

In December 1952 a Conference on Creativity was called together, by invitation, by a sponsoring group from Ohio State University. The artist, the writer, the dancer, the musician were all represented, as well as educators in these various fields. In addition there were those who were interested in the creative process: philosophers, psychiatrists, psychologists. It was a vital and nourishing conference, and led me to produce some rough notes on creativity and the elements which might foster it. These were later expanded into the following paper.

I MAINTAIN that there is a desperate social need for the creative behavior of creative individuals. It is this which justifies the setting forth of a tentative theory of creativity — the nature of the creative act, the conditions under which it occurs, and the manner in which it may constructively be fostered. Such a theory may serve as a stimulus and guide to research studies in this field.

THE SOCIAL NEED

Many of the serious criticisms of our culture and its trends may best be formulated in terms of a dearth of creativity. Let us state some of these very briefly:

In education we tend to turn out conformists, stereotypes, individuals whose education is "completed," rather than freely creative and original thinkers.

In our leisure time activities, passive entertainment and regimented group action are overwhelmingly predominant while creative activities are much less in evidence.

In the sciences, there is an ample supply of technicians, but the number who can creatively formulate fruitful hypotheses and theories is small indeed.

In industry, creation is reserved for the few — the manager, the designer, the head of the research department — while for the many life is devoid of original or creative endeavor.

In individual and family life the same picture holds true. In the clothes we wear, the food we eat, the books we read, and the ideas we hold, there is a strong tendency toward conformity, toward stereotypy. To be original, or different, is felt to be "dangerous."

Why be concerned over this? If, as a people, we enjoy conformity rather than creativity, shall we not be permitted this choice? In my estimation such a choice would be entirely reasonable were it not for one great shadow which hangs over all of us. In a time when knowledge, constructive and destructive, is advancing by the most incredible leaps and bounds into a fantastic atomic age, genuinely creative adaptation seems to represent the only possibility that man can keep abreast of the kaleidoscopic change in his world. With scientific discovery and invention proceeding, we are told, at the rate of geometric progression, a generally passive and culture-bound people cannot cope with the multiplying issues and problems. Unless individuals, groups, and nations can imagine, construct, and creatively revise new ways of relating to these complex changes, the lights will go out. Unless man can make new and original adaptations to his environment as rapidly as his science can change the environment, our culture will perish. Not only individual maladjustment and group tensions, but international annihilation will be the price we pay for a lack of creativity.

Consequently it would seem to me that investigations of the process of creativity, the conditions under which this process occurs, and the ways in which it may be facilitated, are of the utmost importance.

It is in the hope of suggesting a conceptual structure under which such investigations might go forward, that the following sections are offered.

The Creative Process

There are various ways of defining creativity. In order to make more clear the meaning of what is to follow, let me present the elements which, for me, are a part of the creative process, and then attempt a definition.

In the first place, for me as scientist, there must be something observable, some product of creation. Though my fantasies may be extremely novel, they cannot usefully be defined as creative unless they eventuate in some observable product — unless they are symbolized in words, or written in a poem, or translated into a work of art, or fashioned into an invention.

These products must be novel constructions. This novelty grows out of the unique qualities of the individual in his interaction with the materials of experience. Creativity always has the stamp of the individual upon its product, but the product is not the individual, nor his materials, but partakes of the relationship between the two.

Creativity is not, in my judgment, restricted to some particular content. I am assuming that there is no fundamental difference in the creative process as it is evidenced in painting a picture, composing a symphony, devising new instruments of killing, developing a scientific theory, discovering new procedures in human relationships, or creating new formings of one's own personality as in psychotherapy. (Indeed it is my experience in this last field, rather than in one of the arts, which has given me special interest in creativity and its facilitation. Intimate knowledge of the way in which the individual remolds himself in the therapeutic relationship, with originality and effective skill, gives one confidence in the creative potential of all individuals.)

My definition, then, of the creative process is that it is the emergence in action of a novel relational product, growing out of the uniqueness of the individual on the one hand, and the materials, events, people, or circumstances of his life on the other.

Let me append some negative footnotes to this definition. It makes no distinction between "good" and "bad" creativity. One man may be discovering a way of relieving pain, while another is devising a new and more subtle form of torture for political prisoners. Both these actions seem to me creative, even though their social value is very different. Though I shall comment on these social valuations later, I have avoided putting them in my definition because they are so fluctuating. Galileo and Copernicus made creative discoveries which in their own day were evaluated as blasphemous and wicked, and in our day as basic and constructive. We do not want to cloud our definition with terms which rest in subjectivity.

Another way of looking at this same issue is to note that to be regarded historically as representing creativity, the product must be acceptable to some group at some point of time. This fact is not helpful to our definition, however, both because of the fluctuating valuations already mentioned, and also because many creative products have undoubtedly never been socially noticed, but have disappeared without ever having been evaluated. So this concept of group acceptance is also omitted from our definition.

In addition, it should be pointed out that our definition makes no distinction regarding the degree of creativity, since this too is a value judgment extremely variable in nature. The action of the child inventing a new game with his playmates; Einstein formulating a theory of relativity; the housewife devising a new sauce for the meat; a young author writing his first novel; all of these are, in terms of our definition, creative, and there is no attempt to set them in some order of more or less creative.

THE MOTIVATION FOR CREATIVITY

The mainspring of creativity appears to be the same tendency which we discover so deeply as the curative force in psychotherapy

— man's tendency to actualize himself, to become his potentialities. By this I mean the directional trend which is evident in all organic and human life — the urge to expand, extend, develop, mature — the tendency to express and activate all the capacities of the organism, or the self. This tendency may become deeply buried under layer after layer of encrusted psychological defenses; it may be hidden behind elaborate façades which deny its existence; it is my belief however, based on my experience, that it exists in every individual, and awaits only the proper conditions to be released and expressed. It is this tendency which is the primary motivation for creativity as the organism forms new relationships to the environment in its endeavor most fully to be itself.

Let us now attempt to deal directly with this puzzling issue of the social value of a creative act. Presumably few of us are interested in facilitating creativity which is socially destructive. We do not wish, knowingly, to lend our efforts to developing individuals whose creative genius works itself out in new and better ways of robbing, exploiting, torturing, killing, other individuals; or developing forms of political organization or art forms which lead humanity into paths of physical or psychological self-destruction. Yet how is it possible to make the necessary discriminations such that we may encourage a constructive creativity and not a destructive?

The distinction cannot be made by examining the product. The very essence of the creative is its novelty, and hence we have no standard by which to judge it. Indeed history points up the fact that the more original the product, and the more far-reaching its implications, the more likely it is to be judged by contemporaries as evil. The genuinely significant creation, whether an idea, or a work of art, or a scientific discovery, is most likely to be seen at first as erroneous, bad, or foolish. Later it may be seen as obvious, something self-evident to all. Only still later does it receive its final evaluation as a creative contribution. It seems clear that no contemporary mortal can satisfactorily evaluate a creative product at the time that it is formed, and this statement is increasingly true the greater the novelty of the creation.

Nor is it of any help to examine the purposes of the individual participating in the creative process. Many, perhaps most. of the creations and discoveries which have proved to have great social value, have been motivated by purposes having more to do with personal interest than with social value, while on the other hand history records a somewhat sorry outcome for many of those creations (various Utopias, Prohibition, etc.) which had as their avowed purpose the achievement of the social good. No, we must face the fact that the individual creates primarily because it is satisfying to him, because this behavior is felt to be self-actualizing, and we get nowhere by trying to differentiate "good" and "bad" purposes in the creative process.

Must we then give over any attempt to discriminate between creativity which is potentially constructive, and that which is potentially destructive? I do not believe this pessimistic conclusion is justified. It is here that recent clinical findings from the field of psychotherapy give us hope. It has been found that when the individual is "open" to all of his experience (a phrase which will be defined more fully), then his behavior will be creative, and his creativity may be trusted to be essentially constructive.

The differentiation may be put very briefly as follows. To the extent that the individual is denying to awareness (or repressing, if you prefer that term) large areas of his experience, then his creative formings may be pathological, or socially evil, or both. To the degree that the individual is open to all aspects of his experience, and has available to his awareness all the varied sensings and perceivings which are going on within his organism, then the novel products of his interaction with his environment will tend to be constructive both for himself and others. To illustrate, an individual with paranoid tendencies may creatively develop a most novel theory of the relationship between himself and his environment, seeing evidence for his theory in all sorts of minute clues. His theory has little social value, perhaps because there is an enormous range of experience which this individual cannot permit in his awareness. Socrates, on the other hand, while also regarded as "crazy" by his contemporaries, developed novel ideas which have proven to be socially constructive. Very possibly this was because he was notably nondefensive and open to his experience.

The reasoning behind this will perhaps become more clear in the remaining sections of this paper. Primarily however it is based upon the discovery in psychotherapy, that as the individual becomes more open to, more aware of, all aspects of his experience, he is increasingly likely to act in a manner we would term socialized. If he can be aware of his hostile impulses, but also of his desire for friendship and acceptance; aware of the expectations of his culture, but equally aware of his own purposes; aware of his selfish desires, but also aware of his tender and sensitive concern for another; then he behaves in a fashion which is harmonious, integrated, constructive. The more he is open to his experience, the more his behavior makes it evident that the nature of the human species tends in the direction of constructively social living.

The Inner Conditions of Constructive Creativity

What are the conditions within the individual which are most closely associated with a potentially constructive creative act? I see these as possibilities.

A. Openness to experience: Extensionality. This is the opposite of psychological defensiveness, when to protect the organization of the self, certain experiences are prevented from coming into awareness except in distorted fashion. In a person who is open to experience each stimulus is freely relayed through the nervous system, without being distorted by any process of defensiveness. Whether the stimulus originates in the environment, in the impact of form, color, or sound on the sensory nerves, or whether it originates in the viscera, or as a memory trace in the central nervous system, it is available to awareness. This means that instead of perceiving in predetermined categories ("trees are green," "college education is good," "modern art is silly") the individual is aware of this existential moment as *it* is, thus being alive to many experiences which fall outside the usual categories (*this* tree is lavender; *this* college education is damaging; *this* modern sculpture has a powerful effect on me).

This last suggests another way of describing openness to experience. It means lack of rigidity and permeability of boundaries in concepts, beliefs, perceptions, and hypotheses. It means a tolerance for ambiguity where ambiguity exists. It means the ability to receive much conflicting information without forcing closure upon the situation. It means what the general semanticist calls the "extensional orientation."

This complete openness of awareness to what exists at this moment is, I believe, an important condition of constructive creativity. In an equally intense but more narrowly limited fashion it is no doubt present in all creativity. The deeply maladjusted artist who cannot recognize or be aware of the sources of unhappiness in himself, may nevertheless be sharply and sensitively aware of form and color in his experience. The tyrant (whether on a petty or grand scale) who cannot face the weaknesses in himself may nevertheless be completely alive to and aware of the chinks in the psychological armor of those with whom he deals. Because there is the openness to one phase of experience, creativity is possible; because the openness is *only* to one phase of experience, the product of this creativity may be potentially destructive of social values. The more the individual has available to himself a sensitive awareness of all phases of his experience, the more sure we can be that his creativity will be personally and socially constructive.

B. An internal locus of evaluation. Perhaps the most fundamental condition of creativity is that the source or locus of evaluative judgment is internal. The value of his product is, for the creative person, established not by the praise or criticism of others, but by himself. Have I created something satisfying to *me?* Does it express a part of me — my feeling or my thought, my pain or my ecstasy? These are the only questions which really matter to the creative person, or to any person when he is being creative.

This does not mean that he is oblivious to, or unwilling to be aware of, the judgments of others. It is simply that the basis of evaluation lies within himself, in his own organismic reaction to and appraisal of his product. If to the person it has the "feel" of being "me in action," of being an actualization of potentialities in himself which heretofore have not existed and are now emerging into existence, then it is satisfying and creative, and no outside evaluation can change that fundamental fact.

C. The ability to toy with elements and concepts. Though this is probably less important than A or B, it seems to be a condition of creativity. Associated with the openness and lack of rigidity described under A is the ability to play spontaneously with ideas, colors, shapes, relationships — to juggle elements into impossible juxtapositions, to shape wild hypotheses, to make the given problematic, to express the ridiculous, to translate from one form to another, to transform into improbable equivalents. It is from this spontaneous toying and exploration that there arises the hunch, the creative seeing of life in a new and significant way. It is as though out of the wasteful spawning of thousands of possibilities there emerges one or two evolutionary forms with the qualities which give them a more permanent value.

The Creative Act and Its Concomitants

When these three conditions obtain, constructive creativity will occur. But we cannot expect an accurate description of the creative act, for by its very nature it is indescribable. This is the unknown which we must recognize as unknowable until it occurs. This is the improbable that becomes probable. Only in a very general way can we say that a creative act is the natural behavior of an organism which has a tendency to arise when that organism is open to all of its inner and outer experiencing, and when it is free to try out in flexible fashion all manner of relationships. Out of this multitude of half-formed possibilities the organism, like a great computing machine, selects this one which most effectively meets an inner need, or that one which forms a more effective relationship with the environment, or this other one which discovers a more simple and satisfying order in which life may be perceived.

There is one quality of the creative act which may, however, be described. In almost all the products of creation we note a selectivity, or emphasis, an evidence of discipline, an attempt to bring out the essence. The artist paints surfaces or textures in simplified form, ignoring the minute variations which exist in reality. The scientist formulates a basic law of relationships, brushing aside all the particular events or circumstances which might conceal its naked beauty. The writer selects those words and phrases which give unity to his expression. We may say that this is the influence of the specific person, of the "I." Reality exists in a multiplicity of confusing facts, but "I" bring a structure to my relationship to reality; I have "my" way of perceiving reality, and it is this (unconsciously?) disciplined personal selectivity or abstraction which gives to creative products their esthetic quality.

Though this is as far as we can go in describing any aspect of the creative act, there are certain of its concomitants in the individual which may be mentioned. The first is what we may call the Eureka feeling — "This is *it!*" "I have discovered!" "This is what I wanted to express!"

Another concomitant is the anxiety of separateness. I do not believe that many significantly creative products are formed without the feeling, "I am alone. No one has ever done just this before. I have ventured into territory where no one has been. Perhaps I am foolish, or wrong, or lost, or abnormal."

Still another experience which usually accompanies creativity is the desire to communicate. It is doubtful whether a human being can create, without wishing to share his creation. It is the only way he can assuage the anxiety of separateness and assure himself that he belongs to the group. He may confide his theories only to his private diary. He may put his discoveries in some cryptic code. He may conceal his poems in a locked drawer. He may put away his paintings in a closet. Yet he desires to communicate with a group which will understand him, even if he must imagine such a group. He does not create in order to communicate, but once having created he desires to share this new aspect of himself-in-relation-to-his-environment with others.

CONDITIONS FOSTERING CONSTRUCTIVE CREATIVITY

Thus far I have tried to describe the nature of creativity, to indicate that quality of individual experience which increases the likelihood that creativity will be constructive, to set forth the necessary conditions for the creative act and to state some of its concomitants. But if we are to make progress in meeting the social need which was presented initially, we must know whether constructive creativity can be fostered, and if so, how.

From the very nature of the inner conditions of creativity it is clear that they cannot be forced, but must be permitted to emerge. The farmer cannot make the germ develop and sprout from the seed; he can only supply the nurturing conditions which will permit the seed to develop its own potentialities. So it is with creativity. How can we establish the external conditions which will foster and nourish the internal conditions described above? My experience in psychotherapy leads me to believe that by setting up conditions of psychological safety and freedom, we maximize the likelihood of an emergence of constructive creativity. Let me spell out these conditions in some detail, labelling them as X and Y.

X. *Psychological safety.* This may be established by three associated processes.

1. Accepting the individual as of unconditional worth. Whenever a teacher, parent, therapist, or other person with a facilitating function feels basically that this individual is of worth in his own right and in his own unfolding, no matter what his present condition or behavior, he is fostering creativity. This attitude can probably be genuine only when the teacher, parent, etc., senses the potentialities of the individual and thus is able to have an unconditional faith in him, no matter what his present state.

The effect on the individual as he apprehends this attitude, is to sense a climate of safety. He gradually learns that he can be whatever he is, without sham or façade, since he seems to be regarded as of worth no matter what he does. Hence he has less need of rigidity, can discover what it means to be himself, can try to actualize himself in new and spontaneous ways. He is, in other words, moving toward creativity.

2. Providing a climate in which external evaluation is absent. When we cease to form judgments of the other individual from our own locus of evaluation, we are fostering creativity. For the individual to find himself in an atmosphere where he is not being evaluated, not being measured by some external standard, is enormously freeing. Evaluation is always a threat, always creates a need for defensiveness, always means that some portion of experience must be denied to awareness. If this product is evaluated as good by external standards, then I must not admit my own dislike of it. If what I am doing is bad by external standards, then I must not be aware of the fact that it seems to be me, to be part of myself. But if judgments based on external standards are not being made then

I can be more open to my experience, can recognize my own likings and dislikings, the nature of the materials and of my reaction to them, more sharply and more sensitively. I can begin to recognize the locus of evaluation within myself. Hence I am moving toward creativity.

To allay some possible doubts and fears in the reader, it should be pointed out that to cease evaluating another is not to cease having reactions. It may, as a matter of fact, free one to react. "I don't like your idea" (or painting, or invention, or writing), is not an evaluation, but a reaction. It is subtly but sharply different from a judgment which says, "What you are doing is bad (or good), and this quality is assigned to you from some external source." The first statement permits the individual to maintain his own locus of evaluation. It holds the possibility that I am unable to appreciate something which is actually very good. The second statement, whether it praises or condemns, tends to put the person at the mercy of outside forces. He is being told that he cannot simply ask himself whether this product is a valid expression of himself; he must be concerned with what others think. He is being led away from creativity.

3. Understanding empathically. It is this which provides the ultimate in psychological safety, when added to the other two. If I say that I "accept" you, but know nothing of you, this is a shallow acceptance indeed, and you realize that it may change if I actually come to know you. But if I understand you empathically, see you and what you are feeling and doing from your point of view, enter your private world and see it as it appears to you — and still accept you — then this is safety indeed. In this climate you can permit your real self to emerge, and to express itself in varied and novel formings as it relates to the world. This is a basic fostering of creativity.

Y. *Psychological freedom.* When a teacher, parent, therapist, or other facilitating person permits the individual a complete freedom of symbolic expression, creativity is fostered. This permissiveness gives the individual complete freedom to think, to feel, to be, whatever is most inward within himself. It fosters the openness, and the playful and spontaneous juggling of percepts, concepts, and meanings, which is a part of creativity.

Note that it is complete freedom of *symbolic* expression which is described. To express in behavior all feelings, impulses, and formings may not in all instances be freeing. Behavior may in some instances be limited by society, and this is as it should be. But symbolic expression need not be limited. Thus to destroy a hated object (whether one's mother or a rococo building) by destroying a symbol of it, is freeing. To attack it in reality may create guilt and narrow the psychological freedom which is experienced. (I feel unsure of this paragraph, but it is the best formulation I can give at the moment which seems to square with my experience.)

The permissiveness which is being described is not softness or indulgence or encouragement. It is permission to be *free,* which also means that one is responsible. The individual is as free to be afraid of a new venture as to be eager for it; free to bear the consequences of his mistakes as well as of his achievements. It is this type of freedom responsibly to be oneself which fosters the development of a secure locus of evaluation within oneself, and hence tends to bring about the inner conditions of constructive creativity.

Conclusion

I have endeavored to present an orderly way of thinking about the creative process, in order that some of these ideas might be put to a rigorous and objective test. My justification for formulating this theory, and my reason for hoping that such research may be carried out is that the present development of the physical sciences is making an imperative demand upon us, as individuals and as a culture, for creative behavior in adapting ourselves to our new world if we are to survive.

PART VII

The Behavioral Sciences and the Person

*I feel a deep concern
that the developing behavioral sciences
may be used to control the individual and to rob him
of his personhood. I believe, however, that these
sciences might be used to enhance the person.*

20

The Growing Power of the Behavioral Sciences

*Late in 1955 Professor B. F. Skinner of Harvard invited me to participate in a friendly debate with him at the convention of the American Psychological Association in the fall of 1956. He knew that we held very divergent views as to the use of scientific knowledge in molding or controlling human behavior, and suggested that a debate would serve a useful purpose by clarifying the issue. His own basic point of view he had expressed by deploring the unwillingness of psychologists to use their power. "At the moment psychologists are curiously diffident in assuming control where it is available or in developing it where it is not. In most clinics the emphasis is still upon psychometry, and this is in part due to an unwillingness to assume the responsibility of control. . . . In some curious way we feel compelled to leave the active control of human behavior to those who grasp it for selfish purposes."**

I was in agreement with him that such a discussion would serve a valuable purpose in stirring interest in an important issue. We held the debate in September 1956. It attracted a large and attentive audience, and, as is the way in debates, most of the members doubtless left feeling confirmed in the views they held when they came

* Skinner, B. F., in *Current Trends in Psychology*, edited by Wayne Dennis (University of Pittsburgh Press, 1947), pp. 24–25.

in. The text of the debate was published in Science, *Nov. 30, 1956,* 124, *pp. 1057–1066.*

As I mulled over this experience afterward, my only dissatisfaction lay in the fact that it was a debate. While both Skinner and I had endeavored to avoid argument for argument's sake, the tone was nevertheless of an either-or variety. I felt that the question was far too important to be thought of as an argument between two persons, or a simple black versus white issue. So during the following year I wrote out at greater length, and with, I believe, less argumentativeness, my own perception of the elements in this problem which one day will be seen as a profoundly momentous decision for society. The exposition seemed to fall naturally into two parts, and these constitute the two chapters which follow.

I had no particular plan in mind for the use of these documents when I wrote them. I have however used them as the basis for lectures to the course on "Contemporary Trends" at the University of Wisconsin, and this past year I used them as the basis for a seminar presentation to faculty and students at the California Institute of Technology.

THE SCIENCES WHICH DEAL WITH BEHAVIOR are in an infant state. This cluster of scientific disciplines is usually thought of as including psychology, psychiatry, sociology, social psychology, anthropology, and biology, though sometimes the other social sciences such as economics and political science are included, and mathematics and statistics are very much involved as instrumental disciplines. Though they are all at work trying to understand the behavior of man and animals, and though research in these fields is growing by leaps and bounds, it is still an area in which there is undoubtedly more confusion than solid knowledge. Thoughtful workers in these fields tend to stress the enormity of our scientific ignorance regarding behavior, and the paucity of general laws which have been discovered. They tend to compare the state of this field of scientific endeavor with that of physics, and seeing the relative precision of measurement, accuracy of prediction, and elegance and simplicity of the discovered lawfulness in this latter field, are vividly aware of the newness, the infancy, the immaturity, of the behavioral science field.

Without in any way denying the truthfulness of this picture, I believe it is sometimes stressed to the point where the general public may fail to recognize the other side of the coin. Behavioral science, even though in its infancy, has made mighty strides toward becoming an "if — then" science. By this I mean that it has made striking progress in discerning and discovering lawful relationships such that *if* certain conditions exist, *then* certain behaviors will predictably follow. I believe that too few people are aware of the extent, the breadth, and the depth of the advances which have been made in recent decades in the behavioral sciences. Still fewer seem to be aware of the profound social, educational, political, economic, ethical, and philosophical problems posed by these advances.

I would like in this and the subsequent lecture to accomplish several purposes. First, I would like to sketch, in an impressionistic manner, a picture of the growing ability of the behavioral sciences to understand, predict, and control behavior. Then I should like to point out the serious questions and problems which such achievements pose for us as individuals and as a society. Then I should like to suggest a tentative resolution of these problems which has meaning for me.

The "Know-How" of the Behavioral Sciences

Let us try to obtain some impression of the significance of knowledge in the behavioral sciences by dipping in here and there to take a look at specific studies and their meanings. I have endeavored to choose illustrations which would indicate something of the range of the work being done. I am limited by the scope of my own knowledge, and make no claim that these illustrations represent a truly random sampling of the behavioral sciences. I am sure that the fact that I am a psychologist means that I tend to draw a disproportionate share of examples from that field. I have also tended to select illustrations which emphasize the prediction and potential control of behavior, rather than those whose central significance is simply to increase our understanding of behavior. I am quite aware that in the long run these latter studies may lend themselves even more deeply to prediction and control, but their relevance to such problems is not so immediately evident.

In giving these samplings of our scientific knowledge, I shall state them in simple terms, without the various qualifying elements which are important for rigorous accuracy. Each general statement I shall make is supported by reasonably adequate research, though like all scientific findings each statement is an expression of a given degree of probability, not of some absolute truth. Furthermore each statement is open to modification and correction or even refutation through more exact or more imaginative studies in the future.

PREDICTION OF BEHAVIORS

With these selective factors and qualifications in mind let us first look at some of the achievements in the behavioral sciences in which the element of prediction is prominent. The pattern of each of these can be generalized as follows: "If an individual possesses measurable characteristics *a*, *b*, and *c* then we can predict that there is a high probability that he will exhibit behaviors *x*, *y*, and *z*."

Thus, *we know how to predict, with considerable accuracy, which individuals will be successful college students, successful industrial executives, successful insurance salesmen, and the like.* I will not attempt to document this statement, simply because the documentation would be so extensive. The whole field of aptitude testing, of vocational testing, of personnel selection is involved. Although the specialists in these fields are rightly concerned with the degree of inaccuracy in their predictions, the fact remains that here is a wide area in which the work of the behavioral sciences is accepted by multitudes of hardheaded industries, universities and other organizations. We have come to accept the fact that out of an unknown group the behavioral scientist can select (with a certain margin of error) those persons who will be successful typists, practice teachers, filing clerks, or physicists.

This field is continually expanding. Efforts are being made to determine the characteristics of the creative chemist, for example, as over against the merely successful chemist, and, though without outstanding success, efforts have been and are being made to determine the characteristics which will identify the potentially successful psychiatrist and clinical psychologist. Science is moving steadily forward in its ability to say whether or not you possess the measurable characteristics which are associated with a certain type of occupational activity.

We know how to predict success in schools for military officer candidates, and in combat performance. To select one study in this field, Williams and Leavitt (31) found that they could make satisfactory predictions regarding a Marine's probable success in OCS and in later combat performance by obtaining ratings from his "buddies." They also found that in this instance the man's fellow soldiers were better psychological instruments than were the objective tests they used. There is illustrated here not only the use of certain measures to predict behavior, but a willingness to use those measures, whether conventional or unconventional, which are demonstrated to have predictive power.

We can predict how radical or conservative a potential business executive will be. Whyte (30), in his recent book cites this as one of many examples of tests that are in regular use in industrial corporations. Thus in a group of young executives up for promotion, top management can select those who will exhibit (within a margin of error) whatever degree of conservatism or radicalism is calculated to be for the best welfare of the company. They can also base their selection on knowledge of the degree to which each man has a latent hostility to society, or latent homosexuality, or psychotic tendencies. Tests giving (or purporting to give) such measures are in use by many corporations both for screening purposes in selection of new management personnel, and also for purposes of evaluation of men already in management positions, in order to choose those who will be given greater responsibilities.

We know how to predict which members of an organization will be troublemakers and/or delinquent. A promising young psychologist (10) has devised a short, simple pencil and paper test which has shown a high degree of accuracy in predicting which of the employees hired by a department store will be unreliable, dishonest, or otherwise difficult. He states that it is quite possible to identify, with considerable precision, the potential troublemakers in any organized group. This ability to identify those who will make trouble is, so far as the technical issues are concerned, simply an extension of the knowledge we have of prediction in other fields. From the scientific point of view it is no different from predicting who will be a good typesetter.

We know that a competent clerical worker, using a combination of test scores and actuarial tables, can give a better predictive picture of a person's personality and behavior, than can an experienced clinician. Paul Meehl (18) has shown that we are sufficiently advanced in our development of personality tests, and in information accumulated through these tests, that intuitive skill and broad knowledge, experience, and training, are quite unnecessary in producing accurate personality descriptions. He has shown that in many situations in which personality diagnoses are being made — mental hygiene clinics, veteran's hospitals, psychiatric hospitals, and the like, it is wasteful to use well-trained professional personnel to make personality diagnoses through the giving of tests, interviewing the person and the like. He has shown that a clerk can do it better, with only a minimum and impersonal contact with the patient. First a number of tests would be administered and scored. Then the profile of scores would be looked up in actuarial tables prepared on the basis of hundreds of cases, and an accurate and predictive description of personality would emerge, the clerk simply copying down the combination of characteristics which had been found to be statistically correlated with this configuration of scores.

Meehl is here simply carrying forward to the next logical step the current development of psychological instruments for the measurement, appraisal and evaluation of human characteristics, and the prediction of certain behavior patterns on the basis of those measurements. Indeed, there is no reason why Meehl's clerk could not also be eliminated. With proper coded instructions there is no reason why an electronic computer could not score the tests, analyze the profiles and come up with an even more accurate picture of the person and his predicted behavior than a human clerk.

We can select those persons who are easily persuaded, who will conform to group pressures, or those who will not yield. Two separate but compatible studies (15, 16) show that individuals who exhibit certain dependency themes in their responses to the pictures of the Thematic Apperception Test, or who, on another test, show evidence of feelings of social inadequacy, inhibition of aggression, and depressive tendencies, will be easily persuaded, or will yield to group pressures. These small studies are by no means definitive, but there is every reason to suppose that their basic hypothesis is correct and that these or other more refined measures will accurately predict which members of a group will be easily persuaded, and which will be unyielding even to fairly strong group pressures.

We can predict, from the way individuals perceive the movement of a spot of light in a dark room, whether they tend to be prejudiced or unprejudiced. There has been much study of ethnocentrism, the tendency toward a pervasive and rigid distinction between ingroups and outgroups, with hostility toward outgroups, and a submissive attitude toward, and belief in the rightness of, ingroups. One of the theories which has developed is that the more ethnocentric person is unable to tolerate ambiguity or uncertainty in a situation. Operating on this theory Block and Block (5) had subjects report on the degree of movement they perceived in a dim spot of light in a completely dark room. (Actually no movement occurs, but almost all individuals perceive movement in this situation.) They also gave these same subjects a test of ethnocentrism. It was found, as predicted, that those who, in successive trials, quickly established a norm for the amount of movement they perceived, tended to be more ethnocentric than those whose estimates of movement continued to show variety.

This study was repeated, with slight variation, in Australia (28), and the findings were confirmed and enlarged. It was found that the more ethnocentric individuals were less able to tolerate ambiguity, and saw less movement than the unprejudiced. They also were more dependent on others and when making their estimates in the company of another person, tended to conform to the judgment of that person.

Hence it is not too much to say that by studying the way the individual perceives the movement of a dim light in a dark room, we can tell a good deal about the degree to which he is a rigid, prejudiced, ethnocentric person.

This hodgepodge of illustrations of the ability of the behavioral sciences to predict behavior, and hence to select individuals who will exhibit certain behaviors, may be seen simply as the burgeoning applications of a growing field of science. But what these illustrations suggest can also cause a cold chill of apprehension. The thoughtful person cannot help but recognize that these developments I have described are but the beginning. He cannot fail to see that if more highly developed tools were in the hands of an individual or group, together with the power to use them, the social and philosophical implications are awesome. He can begin to see why a scientist like von Bertalanffy warns, "Besides the menace of physical technology, the dangers of psychological technology are often overlooked" (3).

CONDITIONS FOLLOWED BY SPECIFIED BEHAVIORS IN GROUPS

But before we dwell on this social problem, let us move on to another area of the behavioral sciences, and again take a sampling of illustrative studies. This time let us look at some of the research which shows potentiality for *control* of groups. In this realm we are interested in investigations whose findings are of this pattern: "*If* conditions *a*, *b*, and *c* exist or are established in a group, *then* there is a high probability that these conditions will be followed by behaviors *x*, *y*, and *z*."

We know how to provide conditions in a work group, whether in industry or in education, which will be followed by increased productivity, originality, and morale. Studies by Coch and French (7), by Nagle (19), and by Katz, Macoby, and Morse (17) show in general that when workers in industry participate in planning and in decisions, when supervisors are sensitive to worker attitudes, and when supervision is not suspicious or authoritarian, production and morale increase. Conversely we know how to provide the conditions which lead to low productivity and low morale, since the reverse conditions produce a reverse effect.

We know how to establish, in any group, the conditions of leadership which will be followed by personality development in the members of the group, as well as by increased productivity and origi-

nality, and improved group spirit. In groups as diverse as a brief university workshop and an industrial plant making castings, Gordon (9) and Richard (22) have shown that where the leader or leaders hold attitudes customarily thought of as therapeutic, the results are good. In other words if the leader is acceptant, both of the feelings of group members and of his own feelings; if he is understanding of others in a sensitively empathic way; if he permits and encourages free discussion; if he places responsibility with the group; then there is evidence of personality growth in the members of the group, and the group functions more effectively, with greater creativity and better spirit.

We know how to establish conditions which will result in increased psychological rigidity in members of a group. Beier (2), in a careful study, took two matched groups of students and measured several aspects of their abilities, particularly abstract reasoning. Each of the students in one group was then given an analysis of his personality based upon the Rorschach test. Following this both groups were re-tested as to their abilities. The group which had been given an evaluation of their personalities showed a decrease in flexibility, and a significant decrease in ability to carry on abstract reasoning. They became more rigid, anxious, and disorganized in their thinking, in contrast to the control group.

It would be tempting to note that this evaluation — experienced by the group as somewhat threatening — seems very similar to many evaluations made in our schools and universities under the guise of education. All we are concerned with at the moment is that we do know how to establish the conditions which make for less effective functioning on complex intellectual tasks.

We know a great deal about how to establish conditions which will influence consumer responses and/or public opinion. I think this need not be documented with research studies. I refer you to the advertisements in any magazine, to the beguilements of TV programs and their Trendex ratings, to the firms of public relations experts, and to the upward trend of sales by any corporation which puts on a well-planned series of ads.

We know how to influence the buying behavior of individuals by setting up conditions which provide satisfaction for needs of which they are unconscious, but which we have been able to determine. It has been shown that some women who do not buy instant coffee because of "a dislike for its flavor" actually dislike it at a subconscious level because it is associated with being a poor housekeeper — with laziness and spendthrift qualities (11). This type of study, based on the use of projective techniques and "depth" interviews, has led to sales campaigns built upon appeals to the unconscious motives of the individual — his unknown sexual, aggressive, or dependent desires, or as in this instance, the desire for approval.

These illustrative studies indicate something of our potential ability to influence or control the behavior of groups. If we have the power or authority to establish the necessary conditions, the predicted behaviors will follow. There is no doubt that both the studies and the methods are, at the present time crude, but more refined ones are sure to develop.

CONDITIONS WHICH PRODUCE SPECIFIED EFFECTS IN INDIVIDUALS

Perhaps even more impressive than our knowledge of groups is the knowledge which is accumulating in the behavioral sciences as to the conditions which will be followed by specified types of behavior in the individual. It is the possibility of scientific prediction and control of *individual* behavior which comes closest to the interests of each one of us. Again let us look at scattered bits of this type of knowledge.

We know how to set up the conditions under which many individuals will report as true, judgments which are contrary to the evidence of their senses. They will, for example report that Figure A covers a larger area than Figure B, when the evidence of their senses *plainly* indicates that the reverse is true. Experiments by Asch (1) later refined and improved by Crutchfield (8) show that when a person is led to believe that everyone else in the group sees A as larger than B, then he has a strong tendency to go along with this judgment, and in many instances does so with a real belief in his false report.

Not only can we predict that a certain precentage of individuals will thus yield, and disbelieve their own senses, but Crutchfield has determined the personality attributes of those who will do so, and by selection procedures would be able to choose a group who would almost uniformly give in to these pressures for conformity.

We know how to change the opinions of an individual in a selected direction, without his ever becoming aware of the stimuli which changed his opinion. A static, expressionless portrait of a man was flashed on a screen by Smith, Spence and Klein (27). They requested their subjects to note how the expression of the picture changed. Then they intermittently flashed the word "angry" on the screen, at exposures so brief that the subjects were consciously completely unaware of having seen the word. They tended, however, to see the face as becoming more angry. When the word "happy" was flashed on the screen in similar fashion, the viewers tended to see the face as becoming more happy. Thus they were clearly influenced by stimuli which registered at a subliminal level, stimuli of which the individual was not, and could not be, aware.

We know how to influence psychological moods, attitudes, and behaviors, through drugs. For this illustration we step over into the rapidly developing borderline area between chemistry and psychology. From drugs to keep awake while driving or studying, to so-called "truth serum" which reduces the psychological defenses of the individual, to the chemotherapy now practiced in psychiatric wards, the range and complexity of the growing knowledge in this field is striking. Increasingly there are efforts to find drugs with more specific effects — a drug which will energize the depressive individual, another to calm the excited, and the like. Drugs have reportedly been given to soldiers before a battle to eliminate fear. Trade names for the tranquilizing drugs such as Miltown have already crept into our language, even into our cartoons. While much is still unknown in this field, Dr. Skinner of Harvard states that, "In the not-too-distant future, the motivational and emotional conditions of normal life will probably be maintained in any desired state through the use of drugs" (26). While this seems to be a somewhat exaggerated view, his prediction could be partially justified.

We know how to provide psychological conditions which will produce vivid hallucinations and other abnormal reactions in the thoroughly normal individual in the waking state. This knowledge came about as the unexpected by-product of research at McGill University (4). It was discovered that if all channels of sensory stimulation are cut off or muffled, abnormal reactions follow. If healthy subjects lie motionless, to reduce kinaesthetic stimuli, with eyes shielded by translucent goggles which do not permit perception, with hearing largely stifled by foam rubber pillows as well as by being in a quiet cubicle, and with tactile sensations reduced by cuffs over the hands, then hallucinations and bizarre ideation bearing some resemblance to that of the psychotic occur within forty-eight hours in most subjects. What the results would be if the sensory stifling were continued longer is not known because the experience seemed so potentially dangerous that the investigators were reluctant to continue it.

We know how to use a person's own words to open up whole troubled areas in his experience. Cameron (6) and his associates have taken from recorded therapeutic interviews with a patient, brief statements by the patient which seem significantly related to the underlying dynamics of the case. Such a brief statement is then put on a continuous tape so that it can be played over and over. When the patient hears his own significant words repeated again and again, the effect is very potent. By the time it has been repeated twenty or thirty times the patient often begs to have it stopped. It seems clear that it penetrates the individual's defenses, and opens up the whole psychic area related to the statement. For example, a woman who feels very inadequate and is having marital difficulties, talked about her mother in one interview, saying of her, among other things, "That's what I can't understand — that one could strike at a little child." This recorded sentence was played over and over to her. It made her very uneasy and frightened. It also opened up to her all her feelings about her mother. It helped her to see that "not being able to trust my mother not to hurt me has made me mistrustful of everybody." This is a very simple example of the potency of the method, which can not only be helpful but which can be dangerously disorganizing if it penetrates the defenses too deeply or too rapidly.

We know the attitudes which, if provided by a counselor or a therapist, will be predictably followed by certain constructive per-

sonality and behavior changes in the client. Studies we have completed in recent years in the field of psychotherapy (23, 24, 25, 29) justify this statement. The findings from these studies may be very briefly summarized in the following way.

If the therapist provides a relationship in which he is (a) genuine, internally consistent; (b) acceptant, prizing the client as a person of worth; (c) empathically understanding of the client's private world of feelings and attitudes; then certain changes occur in the client. Some of these changes are; the client becomes (a) more realistic in his self-perceptions; (b) more confident and self-directing; (c) more positively valued by himself; (d) less likely to repress elements of his experience; (e) more mature, socialized and adaptive in his behavior; (f) less upset by stress and quicker to recover from it; (g) more like the healthy, integrated, well-functioning person in his personality structure. These changes do not occur in a control group, and appear to be definitely associated with the client's being in a therapeutic relationship.

We know how to disintegrate a man's personality structure, dissolving his self-confidence, destroying the concept he has of himself, and making him dependent on another. A very careful study by Hinkle and Wolff (13) of methods of Communist interrogation of prisoners, particularly in Communist China, has given us a reasonably accurate picture of the process popularly known as "brainwashing." Their study has shown that no magical nor essentially new methods have been used, but mostly a combination of practices developed by rule of thumb. What is involved is largely a somewhat horrifying reversal of the conditions of psychotherapy briefly noted above. If the individual under suspicion is rejected and isolated for a long time, then his need for a human relationship is greatly intensified. The interrogator exploits this by building a relationship in which he shows mostly non-acceptance, and does all he can to arouse guilt, conflict and anxiety. He is acceptant toward the prisoner only when the prisoner "cooperates" by being willing to view events through the interrogator's eyes. He is completely rejecting of the prisoner's internal frame of reference, or personal perception of events. Gradually, out of his need for more acceptance, the prisoner comes to accept halftruths as being true, until little by little he has given up his own view of himself and of his behavior, and has accepted the viewpoint of his interrogator. He is very much demoralized and disintegrated as a person, and largely the puppet of the interrogator. He is then willing to "confess" that he is an enemy of the state, and has committed all kinds of treasonable acts which either he has not done, or which actually had a very different significance.

In a sense it is misleading to describe these methods as a product of the behavioral sciences. They were developed by the Russian and Chinese police, not by scientists. Yet I include them here since it is very clear that these crude methods could be made decidedly more effective by means of scientific knowledge which we now possess. In short our knowledge of how personality and behavior may be changed can be used constructively or destructively, to build or to destroy persons.

CONDITIONS WHICH PRODUCE SPECIFIED EFFECTS IN ANIMALS

Perhaps I have already given ample evidence of the significant and often frightening power of this young field of science. Yet before we turn to the implications of all this, I should like to push the matter one step further by mentioning a few small bits of the very large amount of knowledge which has accumulated in regard to the behavior of animals. Here my own acquaintance is even more limited, but I would like to mention three suggestive studies and their findings.

We know how to establish the conditions which will cause young ducklings to develop a lasting devotion to, for example, an old shoe. Hess (12) has carried out studies of the phenomenon of "imprinting," first investigated in Europe. He has shown that in mallard ducklings, for example, there are a few crucial hours — from the 13th to the 17th hour after hatching — when the duckling becomes attached to any object to which it may be exposed. The more effort it exerts in following this object, the more intense will be the attachment. Normally of course this results in an attachment to the mother duck, but the duckling can just as easily form an indelible devotion to any goal object — to a decoy duck, to a human being, or, as I have mentioned, to an old shoe. Is there any similar tendency in the human infant? One cannot help but speculate.

We know how to eliminate a strong specific fear in a rat by means of electro-convulsive shock. Hunt and Brady (14) first trained thirsty rats to obtain water by pressing a lever. This they did freely and frequently while in the experimental box. When this habit was well fixed a conditioned fear was established by having a clicker sound for a time before a mildly painful electric shock was administered. After a time the rats responded with strong fear reactions and cessation of all lever pressing whenever the clicker sounded, even though the clicking was not followed by any painful stimulus. This conditioned fear reaction was however almost completely eliminated by a series of electo-convulsive shocks administered to the animals. Following this series of shock treatments the animals showed no fear, and freely pressed the lever, even while the clicker was sounding. The authors interpret their results very cautiously, but the general similarity to shock therapy administered to human beings is obvious.

We know how to train pigeons so that they can direct an explosive missile to a pre-determined target. Skinner's amusing account (26a) of this wartime development is only one of many impressive instances of the possibilities of so-called operant conditioning. He took pigeons and "shaped up" their pecking behavior by rewarding them whenever they came at all close to pecking in the direction of, or at, an object he had preselected. Thus he could take a map of a foreign city, and gradually train pigeons to peck only at that portion which contained some vital industry — an airplane factory, for instance. Or he could train them to peck only at representations of certain types of ship at sea. It was then only a technical matter, though to be sure a complex one, to turn their peckings into guidance for a missile. Housing two or three pigeons in the simulated nose of a missile he was able to show that no matter how it might veer off course the pigeons would bring it back "on target" by their pecking.

In response to what I am sure must be your question, I must say that, No, it was never used in warfare, because of the unexpectedly rapid development of electronic devices. But that it would have worked, there seems little question.

Skinner has been able to train pigeons to play ping pong, for example, and he and his co-workers have been able to develop many complex behaviors in animals which seem "intelligent" and "purposeful." The principle is the same in all instances. The animal is given positive reinforcement — some small reward — for every behavior which is at all in the direction of the purpose selected by the investigator. At first perhaps it is only very gross behaviors which in a general way are in the desired direction. But more and more the behavior is "shaped up" to a refined, exact, specific set of preselected actions. From the vast behavioral repertoire of an organism, those behaviors are reinforced with increasing refinement, which serve the exact purpose of the investigator.

Experiments with human beings are a little less clearcut, but it has been shown that by such operant conditioning (such as a nod of the head by the investigator) one can bring about an increase in the number of plural nouns, or statements of personal opinion, expressed by the subject, without his having any awareness of the reason for this change in his behavior. In Skinner's view much of our behavior is the result of such operant conditioning, often unconscious on the part of both participants. He would like to make it conscious and purposeful, and thus controlling of behavior.

We know how to provide animals with a most satisfying experience consisting entirely of electrical stimulation. Olds (20) has found that he can implant tiny electrodes in the septal area of the brain of laboratory rats. When one of these animals presses a bar in his cage, it causes a minute current to pass through these electrodes. This appears to be such a rewarding experience that the animal goes into an orgy of bar pressing, often until he is exhausted. Whatever the subjective nature of the experience it seems to be so satisfying that the animal prefers it to any other activity. I will not speculate as to whether this procedure might be applied to human beings, nor what, in this case, its consequences would be.

THE GENERAL PICTURE AND ITS IMPLICATIONS

I hope that these numerous specific illustrations will have given concrete meaning to the statement that the behavioral sciences are making rapid strides in the understanding, prediction, and control of behavior. In important ways we know how to select individuals who will exhibit certain behaviors; to establish conditions in groups which will lead to various predictable group behaviors; to establish conditions which, in an individual, will lead to specified behavioral results; and in animals our ability to understand, predict and control goes even further, possibly foreshadowing future steps in relation to man.

If your reaction is the same as mine then you will have found that this picture I have given has its deeply frightening aspects. With all the immaturity of this young science, and its vast ignorance, even its present state of knowledge contains awesome possibilities. Suppose some individual or group had both the knowledge available, and the power to use that knowledge for some purpose. Individuals could be selected who would be leaders and others who would be followers. Persons could be developed, enhanced and facilitated, or they could be weakened and disintegrated. Troublemakers could be discovered and dealt with before they became such. Morale could be improved or lowered. Behavior could be influenced by appeals to motives of which the individual was unconscious. It could be a nightmare of manipulation. Admittedly this is wild fantasy, but it is not an impossible fantasy. Perhaps it makes clear the reason why Robert Oppenheimer, one of the most gifted of our natural scientists, looks out from his own domain of physics, and out of the experiences in that field voices a warning. He says that there are some similarities between physics and psychology, and one of these similarities "is the extent to which our progress will create profound problems of decision in the public domain. The physicists have been quite noisy about their contributions in the last decade. The time may well come — as psychology acquires a sound objective corpus of knowledge about human behavior and feeling — when the powers of control thus made available will pose far graver problems than any the physicists have posed." (21)

Some of you may feel that I have somehow made the problem more serious than it is. You may point out that only a very few of the scientific findings I have mentioned have actually been put to use in any way that significantly affects society, and that for the most part these studies are important to the behavioral scientist but have little practical impact on our culture.

I quite agree with this last point. The behavioral sciences at the present time are at somewhat the same stage as the physical sciences several generations ago. As a rather recent example of what I mean, take the argument which occurred around 1900 as to whether a heavier-than-air machine could fly. The science of aeronautics was not well-developed or precise, so that though there were findings which gave an affirmative answer, other studies could be lined up on the negative side. Most important of all, the public did not believe that this science possessed any validity, or would ever significantly affect the culture. They preferred to use their common sense, which told them that man could not possibly fly in a contraption which was heavier than air.

Contrast the public attitude toward aeronautics at that time with the attitude today. We were told, a few years ago, that science predicted we would launch a satellite into space, an utterly fantastic scheme. But so deeply had the public come to have faith in the natural sciences that not a voice was raised in disbelief. The only question the public asked was, "When?"

There is every reason to believe that the same sequence of events will occur in connection with the behavioral sciences. First the public ignores or views with disbelief; then as it discovers that the findings of a science are more dependable than common sense, it begins to use them; the widespread use of the knowledge of a science creates a tremendous demand, so that men and money and effort are poured into the science; finally the development of the science spirals upward at an ever-increasing rate. It seems highly probable that this sequence will be observed in the behavioral sciences. Consequently even though the findings of these sciences are not widely used today, there is every likelihood that they will be widely used tomorrow.

THE QUESTIONS

We have in the making then a science of enormous potential importance, an instrumentality whose social power will make atomic energy seem feeble by comparison. And there is no doubt that the questions raised by this development will be questions of vital importance for this and coming generations. Let us look at a few of these questions.

How shall we use the power of this new science?

What happens to the individual person in this brave new world?

Who will hold the power to use this new knowledge?
Toward what end or purpose or value will this new type of knowledge be used?

I shall try to make a small beginning in the consideration of these questions in the next lecture.

References

1. Asch, Solomon E. *Social Psychology.* New York: Prentice-Hall, 1952, 450–483.

2. Beier, Ernst G. The effect of induced anxiety on some aspects of intellectual functioning. Ph.D. thesis, Columbia University, 1949.

3. Bertalanffy, L. von. A biologist looks at human nature. *Science Monthly*, 1956, *82*, 33–41.

4. Beston, W. H., Woodburn Heron, and T. H. Scott. Effects of decreased variation in the sensory environment. *Canadian J. Psychol.*, 1954, *8*, 70–76.

5. Block, Jack, and Jeanne Block. An investigation of the relationship between intolerance of ambiguity and ethnocentrism. *J. Personality*, 1951, *19*, 303–311.

6. Cameron, D. E. Psychic driving. *Am. J. Psychiat.*, 1956, *112*, 502–509.

7. Coch, Lester, and J. R. P. French, Jr. Overcoming resistance to change, *Human Relations*, 1948, *1*, 512–532.

8. Crutchfield, Richard S. Conformity and character. *Amer. Psychol.*, 1955, *10*, 191–198.

9. Gordon, Thomas. *Group-Centered Leadership.* Chapters 6 to 11. Boston: Houghton Mifflin Co., 1955.

10. Gough, H. E., and D. R. Peterson. The identification and measurement of predispositional factors in crime and delinquency. *J. Consult. Psychol.*, 1952, *16*, 207–212.

11. Haire, M. Projective techniques in marketing research. *J. Marketing*, April 1950, *14*, 649–656.

12. Hess, E. H. An experimental analysis of imprinting — a form of learning. Unpublished manuscript, 1955.

13. Hinkle, L. E., and H. G. Wolff. Communist interrogation and indoctrination of "Enemies of the State." Analysis of methods used by the Communist State Police. *Arch. Neurol. Psychiat.*, 1956, *20*, 115–174.

14. Hunt, H. F., and J. V. Brady. Some effects of electro-convulsive shock on a conditioned emotional response ("anxiety"). *J. Compar. & Physiol. Psychol.*, 1951, *44*, 88–98.

15. Janis, I. Personality correlates of susceptibility to persuasion. *J. Personality*, 1954, *22*, 504–518.

16. Kagan, J., and P. H. Mussen. Dependency themes on the TAT and group conformity. *J. Consult. Psychol.*, 1956, *20*, 29–32.

17. Katz, D., N. Maccoby, and N. C. Morse. *Productivity, supervision, and morale in an office situation.* Part I. Ann Arbor: Survey Research Center, University of Michigan, 1950.

18. Meehl, P. E. Wanted — a good cookbook. *Amer. Psychol.*, 1956, *11*, 263–272.

19. Nagle, B. F. Productivity, employee attitudes, and supervisory sensitivity. *Personnel Psychol.*, 1954, 7, 219–234.

20. Olds, J. A physiological study of reward. In McClelland, D. C. (Ed.). *Studies in Motivation.* New York: Appleton-Century-Crofts, 1955, 134–143.

21. Oppenheimer, R. Analogy in science. *Amer. Psychol.*, 1956, *11*, 127–135.

22. Richard, James, in *Group-Centered Leadership*, by Thomas Gordon, Chapters 12 and 13. Boston: Houghton Mifflin Co., 1955.

23. Rogers, Carl R. *Client-Centered Therapy*. Boston: Houghton Mifflin Co., 1951.

24. Rogers, Carl R. and Rosalind F. Dymond (Eds.). *Psychotherapy and personality change.* University of Chicago Press, 1954.

25. Seeman, Julius, and Nathaniel J. Raskin. Research perspectives in client-centered therapy, in O. H. Mowrer (Ed.). *Phychotherapy: Theory and Research*, Chapter 9. New York: Ronald Press, 1953.

26. Skinner, B. F. The control of human behavior. *Transactions New York Acad. Science*, Series II, Vol. 17, No. 7, May 1955, 547–551.

aspects of life — marriage, child rearing, ethical conduct, work, play, and artistic endeavor. I shall quote from his writings several times.

There are also some writers of fiction who have seen the significance of the coming influence of the behavioral sciences. Aldous Huxley, in his *Brave New World* (1), has given a horrifying picture of saccharine happiness in a scientifically managed world, against which man eventually revolts. George Orwell, in *1984* (5), has drawn a picture of the world created by dictatorial power, in which the behavioral sciences are used as instruments of absolute control of individuals so that not behavior alone but even thought is controlled.

The writers of science fiction have also played a role in visualizing for us some of the possible developments in a world where behavior and personality are as much the subject of science as chemical compounds or electrical impulses.

I should like to try to present, as well as I can, a simplified picture of the cultural pattern which emerges if we endeavor to shape human life in terms of the behavioral sciences.

There is first of all the recognition, almost the assumption, that scientific knowledge is the power to manipulate. Dr. Skinner says: "We must accept the fact that some kind of control of human affairs is inevitable. We cannot use good sense in human affairs unless someone engages in the design and construction of environmental conditions which affect the behavior of men. Environmental changes have always been the condition for the improvement of cultural patterns, and we can hardly use the more effective methods of science without making changes on a grander scale. . . . Science has turned up dangerous processes and materials before. To use the facts and techniques of a science of man to the fullest extent without making some monstrous mistake will be difficult and obviously perilous. It is no time for self-deception, emotional indulgence, or the assumption of attitudes which are no longer useful." (10, p. 56–57)

The next assumption is that such power to control is to be used. Skinner sees it as being used benevolently, though he recognizes the danger of its being misused. Huxley sees it as being used with benevolent intent, but actually creating a nightmare. Orwell describes the results if such power is used malignantly, to enhance the degree of regulation exercised by a dictatorial government.

26a. ———. Pigeons in a Pelican, *Amer. Psychol.*, 1960, *15*, 28–37.

27. Smith, G. J. W., Spence, D. P., and Klein, G. S., Subliminal effects of verbal stimuli, *Jour. Abn. & Soc. Psychol.*, 1959, *59*, 167–176.

28. Taft, R. Intolerance of ambiguity and ethnocentrism. *J. Consult. Psychol.*, 1956, *20*, 153–154.

29. Thetford, William N. An objective measure of frustration tolerance in evaluating psychotherapy, in W. Wolff (Ed.). *Success in psychotherapy*, Chapter 2. New York: Grune and Stratton, 1952.

30. Whyte, W. H. *The Organization Man.* New York: Simon & Schuster, 1956.

31. Williams, S. B., and H. J. Leavitt. Group opinion as a predictor of military leadership. *J. Consult. Psychol.*, 1947, *11*, 283–291.

21

The Place of the Individual in the New World of the Behavioral Sciences

IN THE PRECEDING LECTURE I endeavored to point out, in a very sketchy manner, the advances of the behavioral sciences in their ability to predict and control behavior. I tried to suggest the new world into which we will be advancing at an evermore headlong pace. Today I want to consider the question of how we — as individuals, as groups, as a culture — will live in, will respond to, will adapt to, this brave new world. What stance will we take in the face of these new developments?

I am going to describe two answers which have been given to this question, and then I wish to suggest some considerations which may lead to a third answer.

Deny and Ignore

One attitude which we can take is to deny that these scientific advances are taking place, and simply take the view that there can be no study of human behavior which is truly scientific. We can hold that the human animal cannot possibly take an objective attitude toward himself, and that therefore no real science of behavior can exist. We can say that man is always a free agent, in some sense that

makes scientific study of his behavior impossible. Not long ago, at a conference on the social sciences, curiously enough, I heard a well known economist take just this view. And one of this country's most noted theologians writes, "In any event, no scientific investigation of past behavior can become the basis of predictions of future behavior." (3, p. 47)

The attitude of the general public is somewhat similar. Without necessarily denying the possibility of a behavioral science, the man in the street simply ignores the developments which are taking place. To be sure he becomes excited for a time when he hears it said that the Communists have attempted to change the soldiers they have captured, by means of "brainwashing." He may show a mild reaction of annoyance to the revelations of a book such as Whyte's (13) which shows how heavily, and in what manipulative fashion, the findings of the behavioral sciences are used by modern industrial corporations. But by and large he sees nothing in all this to be concerned about, any more than he did in the first theoretical statements that the atom could be split.

We may, if we wish, join him in ignoring the problem. We may go further, like the older intellectuals I have cited, and looking at the behavorial sciences may declare that "there ain't no such animal." But since these reactions do not seem particularly intelligent I shall leave them to describe a much more sophisticated and much more prevalent point of view.

The Formulation of Human Life in Terms of Science

Among behavioral scientists it seems to be largely taken for granted that the findings of such science will be used in the prediction and control of human behavior. Yet most psychologists and other scientists have given little thought to what this would mean. An exception to this general tendency is Dr. B. F. Skinner of Harvard who has been quite explicit in urging psychologists to use the powers of control which they have in the interest of creating a better world. In an attempt to show what he means Dr. Skinner wrote a book some years ago entitled *Walden Two* (12), in which he gives a fictional account of what he regards as a Utopian community in which the learnings of the behavioral sciences are fully utilized in all

STEPS IN THE PROCESS

Let us look at some of the elements which are involved in the concept of the control of human behavior as mediated by the behavioral sciences. What would be the steps in the process by which a society might organize itself so as to formulate human life in terms of the science of man?

First would come the selection of goals. In a recent paper Dr. Skinner suggests that one possible goal to be assigned to the behavioral technology is this: "Let man be happy, informed, skillful, well-behaved, and productive" (10, p. 47). In his *Walden Two*, where he can use the guise of fiction to express his views, he becomes more expansive. His hero says, "Well, what do you say to the design of personalities? Would that interest you? The control of temperament? Give me the specifications, and I'll give you the man! What do you say to the control of motivation, building the interests which will make men most productive and most successful? Does that seem to you fantastic? Yet some of the techniques are available, and more can be worked out experimentally. Think of the possibilities! . . . Let us control the lives of our children and see what we can make of them." (12, p. 243)

What Skinner is essentially saying here is that the current knowledge in the behavioral sciences plus that which the future will bring, will enable us to specify, to a degree which today would seem incredible, the kind of behavioral and personality results which we wish to achieve. This is obviously both an opportunity and a very heavy burden.

The second element in this process would be one which is familiar to every scientist who has worked in the field of applied science. Given the purpose, the goal, we proceed by the method of science — by controlled experimentation — to discover the means to these ends. If for example our present knowledge of the conditions which cause men to be productive is limited, further investigation and experimentation would surely lead us to new knowledge in this field. And still further work will provide us with the knowledge of even more effective means. The method of science is self-correcting in thus arriving at increasingly effective ways of achieving the purpose we have selected.

The third element in the control of human behavior through the behavioral sciences involves the question of power. As the conditions or methods are discovered by which to achieve our goal, some person or group obtains the power to establish those conditions or use those methods. There has been too little recognition of the problem involved in this. To hope that the power being made available by the behavioral sciences will be exercised by the scientists, or by a benevolent group, seems to me a hope little supported by either recent or distant history. It seems far more likely that behavioral scientists, holding their present attitudes, will be in the position of the German rocket scientists specializing in guided missiles. First they worked devotedly for Hitler to destroy Russia and the United States. Now depending on who captured them, they work devotedly for Russia in the interest of destroying the United States, or devotedly for the United States in the interest of destroying Russia. If behavioral scientists are concerned solely with advancing their science, it seems most probable that they will serve the purposes of whatever individual or group has the power.

But this is, in a sense a digression. The main point of this view is that some person or group will have and use the power to put into effect the methods which have been discovered for achieving the desired goal.

The fourth step in this process whereby a society might formulate its life in terms of the behavioral sciences is the exposure of individuals to the methods and conditions mentioned. As individuals are exposed to the prescribed conditions this leads, with a high degree of probability, to the behavior which has been desired. Men then become productive, if that has been the goal, or submissive, or whatever it has been decided to make them.

To give something of the flavor of this aspect of the process as seen by one of its advocates, let me again quote the hero of *Walden Two*. "Now that we *know* how positive reinforcement works, and why negative doesn't" he says, commenting on the method he is advocating, "we can be more deliberate and hence more successful, in our cultural design. We can achieve a sort of control under which the controlled, though they are following a code much more scrupulously than was ever the case under the old system, nevertheless *feel free.* They are doing what they want to do, not what they are forced to do. That's the source of the tremendous power of positive reinforcement — there's no restraint and no revolt. By a careful design, we control not the final behavior, but the *inclination* to behave — the motives, the desires, the wishes. The curious thing is that in that case *the question of freedom never arises.*" (12, p. 218)

The Picture and It's Implications

Let me see if I can sum up very briefly the picture of the impact of the behavioral sciences upon the individual and upon society, as this impact is explicitly seen by Dr. Skinner, and implied in the attitudes and work of many, perhaps most, behavioral scientists. Behavioral science is clearly moving forward; the increasing power for control which it gives will be held by some one or some group; such an individual or group will surely choose the purposes or goals to be achieved; and most of us will then be increasingly controlled by means so subtle we will not even be aware of them as controls. Thus whether a council of wise psychologists (if this is not a contradiction in terms) or a Stalin or a Big Brother has the power, and whether the goal is happiness, or productivity, or resolution of the Oedipus complex, or submission, or love of Big Brother, we will inevitably find ourselves moving toward the chosen goal, and probably thinking that we ourselves desire it. Thus if this line of reasoning is correct, it appears that some form of completely controlled society — a *Walden Two* or a *1984* — is coming. The fact that is would surely arrive piecemeal rather than all at once, does not greatly change the fundamental issues. Man and his behavior would become a planned product of a scientific society.

You may well ask, "But what about individual freedom? What about the democratic concepts of the rights of the individual?" Here too Dr. Skinner is quite specific. He says quite bluntly. "The hypothesis that man is not free is essential to the application of scientific method to the study of human behavior. The free inner man who is held responsible for the behavior of the external biological organism is only a pre-scientific substitute for the kinds of causes which are discovered in the course of a scientific analysis. All these alternative causes lie *outside* the individual." (11, p. 447)

In another source he explains this at somewhat more length. "As the use of science increases, we are forced to accept the theoretical structure with which science represents its facts. The difficulty is that this structure is clearly at odds with the traditional democratic conception of man. Every discovery of an event which has a part in shaping a man's behavior seems to leave so much the less to be credited to the man himself; and as such explanations become more and more comprehensive, the contribution which may be claimed by the individual himself appears to approach zero. Man's vaunted creative powers, his original accomplishments in art, science and morals, his capacity to choose and our right to hold him responsible for the consequences of his choice — none of these is conspicuous in this new self-portrait. Man, we once believed, was free to express himself in art, music and literature, to inquire into nature, to seek salvation in his own way. He could initiate action and make spontaneous and capricious changes of course. Under the most extreme duress some sort of choice remained to him. He could resist any effort to control him, though it might cost him his life. But science insists that action is initiated by forces impinging upon the individual, and that caprice is only another name for behavior for which we have not yet found a cause." (10, p. 52–53)

The democratic philosophy of human nature and of government is seen by Skinner as having served a useful purpose at one time. "In rallying men against tyranny it was necessary that the individual be strengthened, that he be taught that he had rights and could govern himself. To give the common man a new conception of his worth, his dignity, and his power to save himself, both here and hereafter, was often the only resource of the revolutionist." (10, p. 53) He regards this philosophy as being now out of date and indeed an obstacle "if it prevents us from applying to human affairs the science of man." (10, p. 54)

A Personal Reaction

I have endeavored, up to this point, to give an objective picture of some of the developments in the behavioral sciences, and an objective picture of the kind of society which might emerge out of these developments. I do however have strong personal reactions to the kind of world I have been describing, a world which Skinner explicitly (and many other scientists implicitly) expect and hope for in the future. To me this kind of world would destroy the human person as I have come to know him in the deepest moments of psychotherapy. In such moments I am in relationship with a person who is spontaneous, who is responsibly free, that is, aware of this freedom to choose who he will be, and aware also of the consequences of his choice. To believe, as Skinner holds, that all this is an illusion, and that spontaneity, freedom, responsibility, and choice have no real existence, would be impossible for me.

I feel that to the limit of my ability I have played my part in advancing the behavioral sciences, but if the result of my efforts and those of others is that man becomes a robot, created and controlled by a science of his own making, then I am very unhappy indeed. If the good life of the future consists in so conditioning individuals through the control of their environment, and through the control of the rewards they receive, that they will be inexorably productive, well-behaved, happy or whatever, then I want none of it. To me this is a pseudo-form of the good life which includes everything save that which makes it good.

And so I ask myself, is there any flaw in the logic of this development? Is there any alternative view as to what the behavioral sciences might mean to the individual and to society? It seems to me that I perceive such a flaw, and that I can conceive of an alternative view. These I would like to set before you.

Ends and Values in Relation to Science

It seems to me that the view I have presented rests upon a faulty perception of the relationship of goals and values to the enterprise of science. The significance of the *purpose* of a scientific undertaking is, I believe, grossly underestimated. I would like to state a two-pronged thesis which in my estimation deserves consideration. Then I will elaborate the meaning of these two points.

1. In any scientific endeavor — whether "pure" or applied science — there is a prior personal subjective choice of the purpose or value which that scientific work is perceived as serving.

2. This subjective value choice which brings the scientific en-

deavor into being must always lie outside of that endeavor, and can never become a part of the science involved in that endeavor.

Let me illustrate the first point from Dr. Skinner's writings. When he suggests that the task for the behavioral sciences is to make man "productive," "well-behaved," etc., it is obvious that he is making a choice. He might have chosen to make men submissive, dependent, and gregarious, for example. Yet by his own statement in another context man's "capacity to choose," his freedom to select his course and to initiate action — these powers do not exist in the scientific picture of man. Here is, I believe, the deep-seated contradiction, or paradox. Let me spell it out as clearly as I can.

Science, to be sure, rests on the assumption that behavior is caused — that a specified event is followed by a consequent event. Hence all is determined, nothing is free, choice is impossible. But we must recall that science itself, and each specific scientific endeavor, each change of course in a scientific research, each interpretation of the meaning of a scientific finding and each decision as to how the finding shall be applied, rests upon a personal subjective choice. Thus science in general exists in the same paradoxical situation as does Dr. Skinner. A personal subjective choice made by man sets in motion the operations of science, which in time proclaims that there can be no such thing as a personal subjective choice. I shall make some comments about this continuing paradox at a later point.

I stressed the fact that each of these choices initiating or furthering the scientific venture, is a value choice. The scientist investigates this rather than that, because he feels the first investigation has more value for him. He chooses one method for his study rather than another because he values it more highly. He interprets his findings in one way rather than another because he believes the first way is closer to the truth, or more valid — in other words that it is closer to a criterion which he values. Now these value choices are never a part of the scientific venture itself. The value choices connected with a particular scientific enterprise always and necessarily lie outside of that enterprise.

I wish to make it clear that I am not saying that values cannot be included as a subject of science. It is not true that science deals only with certain classes of "facts" and that these classes do not include values. It is a bit more complex than that, as a simple illustration or two may make clear.

If I value knowledge of the "three R's" as a goal of education, the methods of science can give me increasingly accurate information as to how this goal may be achieved. If I value problem-solving ability as a goal of education, the scientific method can give me the same kind of help.

Now if I wish to determine whether problem-solving ability is "better" than knowledge of the three R's, then scientific method can also study those two values, but *only* — and this is very important — only in terms of some other value which I have subjectively chosen. I may value college success. Then I can determine whether problem-solving ability or knowledge of the three R's is most closely associated with that value. I may value personal integration or vocational success or responsible citizenship. I can determine whether problem-solving ability or knowledge of the three R's is "better" for achieving any one of these values. But the value or purpose which gives meaning to a particular scientific endeavor must always lie outside of that endeavor.

Though our concern in these lectures is largely with applied science what I have been saying seems equally true of so-called pure science. In pure science the usual prior subjective value choice is the discovery of truth. But this is a subjective choice, and science can never say whether it is the best choice, save in the light of some other value. Geneticists in Russia, for example, had to make a subjective choice of whether it was better to pursue truth, or to discover facts which upheld a governmental dogma. Which choice is "better"? We could make a scientific investigation of those alternatives, but only in the light of some other subjectively chosen value. If, for example, we value the survival of a culture then we could begin to investigate with the methods of science the question as to whether pursuit of truth or support of governmental dogma is most closely associated with cultural survival.

My point then is that any scientific endeavor, pure or applied, is carried on in the pursuit of a purpose or value which is subjectively chosen by persons. It is important that this choice be made explicit, since the particular value which is being sought can never be tested or evaluated, confirmed or denied, by the scientific endeavor to which it gives birth and meaning. The initial purpose or value always and necessarily lies outside the scope of the scientific effort which it sets in motion.

Among other things this means that if we choose some particular goal or series of goals for human beings, and then set out on a large scale to control human behavior to the end of achieving those goals, we are locked in the rigidity of our initial choice, because such a scientific endeavor can never transcend itself to select new goals. Only subjective human persons can do that. Thus if we choose as our goal the state of happiness for human beings (a goal deservedly ridiculed by Aldous Huxley in *Brave New World*), and if we involved all of society in a successful scientific program by which people became happy, we would be locked in a colossal rigidity in which no one would be free to question this goal, because our scientific operations could not transcend themselves to question their guiding purposes. And without laboring this point, I would remark that colossal rigidity, whether in dinosaurs or dictatorships, has a very poor record of evolutionary survival.

If, however, a part of our scheme is to set free some "planners" who do not have to be happy, who are not controlled, and who are therefore free to choose other values, this has several meanings. It means that the purpose we have chosen as our goal is not a sufficient and satisfying one for human beings, but must be supplemented. It also means that if it is necessary to set up an elite group which is free, then this shows all too clearly that the great majority are only the slaves — no matter by what high-sounding name we call them — of those who select the goals.

Perhaps, however, the thought is that a continuing scientific endeavor will evolve its own goals; that the initial findings will alter the directions, and subsequent findings will alter them still further and that the science somehow develops its own purpose. This seems to be a view implicitly held by many scientists. It is surely a reasonable description, but it overlooks one element in this continuing development, which is that subjective personal choice enters in at every point at which the direction changes. The findings of a science, the results of an experiment, do not and never can tell us what next scientific purpose to pursue. Even in the purest of science, the scientist must decide what the findings mean, and must subjectively choose what next step will be most profitable in the pursuit of his purpose. And if we are speaking of the application of scientific knowledge, then it is distressingly clear that the increasing scientific knowledge of the structure of the atom carries with it no necessary choice as to the purpose to which this knowledge will be put. This is a subjective personal choice which must be made by many individuals.

Thus I return to the proposition with which I began this section of my remarks — and which I now repeat in different words. Science has its meaning as the objective pursuit of a purpose which has been subjectively chosen by a person or persons. This purpose or value can never be investigated by the particular scientific experiment or investigation to which it has given birth and meaning. Consequently, any discussion of the control of human beings by the behavioral sciences must first and most deeply concern itself with the subjectively chosen purposes which such an application of science is intended to implement.

An Alternative Set of Values

If the line of reasoning I have been presenting is valid, then it opens new doors to us. If we frankly face the fact that science takes off from a subjectively chosen set of values, then we are free to select the values we wish to pursue. We are not limited to such stultifying goals as producing a controlled state of happiness, productivity, and the like. I would like to suggest a radically different alternative.

Suppose we start with a set of ends, values, purposes, quite different from the type of goals we have been considering. Suppose we do this quite openly, setting them forth as a possible value choice to be accepted or rejected. Suppose we select a set of values which focuses on fluid elements of process, rather than static attributes. We might then value:

Man as a process of becoming; as a process of achieving worth and dignity through the development of his potentialities;

The individual human being as a self-actualizing process, moving on to more challenging and enriching experiences;

The process by which the individual creatively adapts to an ever-new and changing world;

The process by which knowledge transcends itself, as for example the theory of relativity transcended Newtonian physics, itself to be transcended in some future day by a new perception.

If we select values such as these we turn to our science and technology of behavior with a very different set of questions. We will want to know such things as these:

Can science aid us in the discovery of new modes of richly rewarding living? More meaningful and satisfying modes of interpersonal relationships?

Can science inform us as to how the human race can become a more intelligent participant in its own evolution — its physical, psychological and social evolution?

Can science inform us as to ways of releasing the creative capacity of individuals, which seem so necessary if we are to survive in this fantastically expanding atomic age? Dr. Oppenheimer has pointed out (4) that knowledge, which used to double in millenia or centuries, now doubles in a generation or a decade. It appears that we will need to discover the utmost in release of creativity if we are to be able to adapt effectively.

In short, can science discover the methods by which man can most readily become a continually developing and self-transcending process, in his behavior, his thinking, his knowledge? Can science predict and release an essentially "unpredictable" freedom?

It is one of the virtues of science as a method that it is as able to advance and implement goals and purposes of this sort as it is to serve static values such as states of being well-informed, happy, obedient. Indeed we have some evidence of this.

A Small Example

I will perhaps be forgiven if I document some of the possibilities along this line by turning to psychotherapy, the field I know best.

Psychotherapy, as Meerloo (2) and others have pointed out, can be one of the most subtle tools for the control of one person by another. The therapist can subtly mold individuals in imitation of himself. He can cause an individual to become a submissive and conforming being. When certain therapeutic principles are used in extreme fashion, we call it brainwashing, an instance of the disintegration of the personality and a reformulation of the person along lines desired by the controlling individual. So the principles of therapy can be used as a most effective means of external control of human personality and behavior. Can psychotherapy be anything else?

Here I find the developments going on in client-centered psychotherapy (8) an exciting hint of what a behavioral science can do in achieving the kinds of values I have stated. Quite aside from being a somewhat new orientation in psychotherapy, this development has important implications regarding the relation of a behavioral science to the control of human behavior. Let me describe our experience as it relates to the issues of the present discussion.

In client-centered therapy, we are deeply engaged in the prediction and influencing of behavior. As therapists we institute certain attitudinal conditions, and the client has relatively little voice in the establishment of these conditions. Very briefly we have found that the therapist is most effective if he is: (a) genuine, integrated, transparently real in the relationship; (b) acceptant of the client as a separate, different, person, and acceptant of each fluctuating aspect of the client as it comes to expression; and (c) sensitively empathic in his understanding, seeing the world through the client's eyes. Our research permits us to predict that if these attitudinal conditions are instituted or established, certain behavioral consequences will ensue. Putting it this way sounds as if we are again back in the familiar groove of being able to predict behavior, and hence able to control it. But precisely here exists a sharp difference.

The conditions we have chosen to establish predict such behavioral consequences as these: that the client will become more self-directing, less rigid, more open to the evidence of his senses, better organized and integrated, more similar to the ideal which he has chosen for himself. In other words we have established by external control conditions which we predict will be followed by internal control by the individual, in pursuit of internally chosen goals. We have set the conditions which predict various classes of behaviors — self-directing behaviors, sensitivity to realities within and without, flexible adaptiveness — which are by their very nature *unpredictable* in their specifics. The conditions we have established predict behavior which is essentially "free." Our recent research (9) indicates that our predictions are to a significant degree corroborated, and our commitment to the scientific method causes us to believe that more effective means of achieving these goals may be realized.

Research exists in other fields — industry, education, group dynamics — which seems to support our own findings. I believe it may be conservatively stated that scientific progress has been made in identifying those conditions in an interpersonal relationship which, if they exist in B, are followed in A by greater maturity in behavior, less dependence upon others, an increase in expressiveness as a person, an increase in variability, flexibility and effectiveness of adaptation, an increase in self-responsibility and self-direction. And quite in contrast to the concern expressed by some we do not find that the creatively adaptive behavior which results from such self-directed variability of expression is too chaotic or too fluid. Rather, the individual who is open to his experience, and self-directing, is harmonious, not chaotic, ingenious rather than random, as he orders his responses imaginatively toward the achievement of his own purposes. His creative actions are no more a chaotic accident than was Einstein's development of the theory of relativity.

Thus we find ourselves in fundamental agreement with John Dewey's statement: "Science has made its way by releasing, not by suppressing, the elements of variation, of invention and innovation, of novel creation in individuals." (7, p. 359) We have come to believe that progress in personal life and in group living is made in the same way, by releasing variation, freedom, creativity.

A Possible Concept of the Control of Human Behavior

It is quite clear that the point of view I am expressing is in sharp contrast to the usual conception of the relationship of the behavioral sciences to the control of human behavior, previously mentioned. In order to make this contrast even more blunt, I will state this possibility in a form parallel to the steps which I described before.

1. It is possible for us to choose to value man as a self-actualizing process of becoming; to value creativity, and the process by which knowledge becomes self-transcending.

2. We can proceed, by the methods of science, to discover the conditions which necessarily precede these processes, and through continuing experimentation, to discover better means of achieving these purposes.

3. It is possible for individuals or groups to set these conditions, with a minimum of power or control. According to present knowledge, the only authority necessary is the authority to establish certain qualities of interpersonal relationship.

4. Exposed to these conditions, present knowledge suggests that individuals become more self-responsible, make progress in self-actualization, become more flexible, more unique and varied, more creatively adaptive.

5. Thus such an initial choice would inaugurate the beginnings of a social system or subsystem in which values, knowledge, adaptive skills, and even the concept of science would be continually changing and self-transcending. The emphasis would be upon man as a process of becoming.

I believe it is clear that such a view as I have been describing does not lead to any definable Utopia. It would be impossible to predict its final outcome. It involves a step by step development, based upon a continuing subjective choice of purposes, which are implemented by the behavioral sciences. It is in the direction of the "open society," as that term has been defined by Popper (6), where individuals carry responsibility for personal decisions. It is at the opposite pole from his concept of the closed society, of which *Walden Two* would be an example.

I trust it is also evident that the whole emphasis is upon process, not upon end states of being. I am suggesting that it is by choosing to value certain qualitative elements of the process of becoming, that we can find a pathway toward the open society.

The Choice

It is my hope that I have helped to clarify the range of choice which will lie before us and our children in regard to the behavioral sciences. We can choose to use our growing knowledge to enslave people in ways never dreamed of before, depersonalizing them, controlling them by means so carefully selected that they will perhaps never be aware of their loss of personhood. We can choose to utilize our scientific knowledge to make men necessarily happy, well-behaved, and productive, as Dr. Skinner suggests. We can, if we wish, choose to make men submissive, conforming, docile. Or at the other end of the spectrum of choice we can choose to use the behavioral sciences in ways which will free, not control; which will bring about constructive variability, not conformity; which will develop creativity, not contentment; which will facilitate each person in his self-directed process of becoming; which will aid individuals, groups, and even the concept of science, to become self-transcending in freshly adaptive ways of meeting life and its problems. The choice is up to us, and the human race being what it is, we are likely to stumble about, making at times some nearly disastrous value choices, and at other times highly constructive ones.

If we choose to utilize our scientific knowledge to free men, then it will demand that we live openly and frankly with the great paradox of the behavioral sciences. We will recognize that behavior, when examined scientifically, is surely best understood as determined by prior causation. This is the great fact of science. But responsible personal choice, which is the most essential element in being a person, which is the core experience in psychotherapy, which exists prior to any scientific endeavor, is an equally prominent fact in our lives. We will have to live with the realization that to deny the reality of the experience of responsible personal choice is as stultifying, as closed-minded, as to deny the possibility of a behavioral science. That these two important elements of our experience appear to be in contradiction has perhaps the same significance as the contradiction between the wave theory and the corpuscular theory of light, both of which can be shown to be true, even though incompatible. We cannot profitably deny our subjective life, any more than we can deny the objective description of that life.

In conclusion then, it is my contention that science cannot come into being without a personal choice of the values we wish to achieve. And these values we choose to implement will forever lie outside of the science which implements them; the goals we select, the purposes we wish to follow, must always be outside of the science which achieves them. To me this has the encouraging meaning that the human person, with his capacity of subjective choice, can and will always exist, separate from and prior to any of his scientific undertakings. Unless as individuals and groups we choose to relinquish our capacity of subjective choice, we will always remain free persons, not simply pawns of a self-created behavioral science.

References

1. Huxley, A. *Brave New World.* New York and London: Harper and Bros., 1946.
2. Meerloo, J. A. M. Medication into submission: the danger of therapeutic coercion. *J. Nerv. Ment. Dis.*, 1955, *122*, 353–360.
3. Niebuhr, R. *The Self and the Dramas of History.* New York: Scribner, 1955.
4. Oppenheimer, R. Science and our times. *Roosevelt University Occasional Papers.* 1956, 2, Chicago, Illinois.
5. Orwell, G. *1984.* New York: Harcourt, Brace, 1949; New American Library, 1953.
6. Popper, K. R. *The Open Society and Its Enemies.* London: Routledge and Kegan Paul, 1945.
7. Ratner, J. (Ed.). *Intelligence in the Modern World: John Dewey's Philosophy.* New York: Modern Library, 1939.
8. Rogers, C. R. *Client-Centered Therapy.* Boston: Houghton Mifflin, 1951.
9. Rogers, C. R. and Rosalind Dymond (Eds.). *Psychotherapy and Personality Change.* University of Chicago Press, 1954.
10. Skinner, B. F. Freedom and the control of men. *Amer. Scholar*, Winter, 1955–56, *25*, 47–65.

11. Skinner, B. F. *Science and Human Behavior.* New York: Macmillan, 1953. Quotation by permission of The Macmillan Co.

12. Skinner, B. F. *Walden Two.* New York: Macmillan 1948. Quotations by permission of The Macmillan Co.

13. Whyte, W. H. *The Organization Man.* New York: Simon & Schuster, 1956.

Appendix

Chronological Bibliography

PUBLICATIONS OF CARL R. ROGERS
1930–1960 *inclusive*

1930

With C. W. Carson. Intelligence as a factor in camping activities. *Camping Magazine*, 1930, *3* (3), 8–11.

1931

Measuring Personality Adjustment in Children Nine to Thirteen. New York: Teachers College, Columbia University, Bureau of Publications, 1931, 107 pp.

A Test of Personality Adjustment. New York: Association Press, 1931.

With M. E. Rappaport. We pay for the Smiths. *Survey Graphic*, 1931, *19*, 508 ff.

1933

A good foster home: Its achievements and limitations. *Mental Hygiene*, 1933, *17*, 21–40. Also published in F. Lowry (Ed.), *Readings in Social Case Work.* Columbia University Press, 1939, 417–436.

1936

Social workers and legislation. *Quarterly Bulletin New York State Conference on Social Work*, 7 (3), 1936, 3–9.

1937

Three surveys of treatment measures used with children. *Amer. J. Orthopsychiat.*, 1937, 7, 48–57.

The clinical psychologist's approach to personality problems. *The Family*, 1947, *18*, 233–243.

1938

A diagnostic study of Rochester youth. *N. Y. State Conference on Social Work.* Syracuse: 1938, 48–54.

1939

The Clinical Treatment of the Problem Child. Boston: Houghton Mifflin, 1939, 393 pp.

Needed emphases in the training of clinical psychologists. *J. Consult. Psychol.*, 1939, *3*, 141–143.

Authority and case work — are they compatible? *Quarterly Bulletin, N. Y. State Conference on Social Work.* Albany: 1939, 16–24.

1940

The processes of therapy. *J. Consult, Psychol.*, 1940, *4*, 161–164.

1941

Psychology in clinical practice. In J. S. Gray (Ed.), *Psychology in Use.* New York: American Book Company, 1941, 114–167.

With C. C. Bennett. Predicting the outcomes of treatment. *Amer. J. Orthopsychiat.*, 1941, *11*, 210–221.

With C. C. Bennett. The clinical significance of problem syndromes. *Amer. J. Orthopsychiat.*, 1941, *11*, 222–229.

1942

The psychologist's contributions to parent, child, and community problems. *J. Consult. Psychol.*, 1942, *6*, 8–18.

A study of the mental health problems in three representative elementary schools. In T. C. Holy *et al.*, *A Study of Health and Physical Education in Columbus Public Schools.* Ohio State Univer., Bur. of Educ. Res. Monogr., No. 25, 1942, 130–161.

Mental health problems in three elementary schools. *Educ. Research Bulletin*, 1942, *21*, 69–79.

The use of electrically recorded interviews in improving psychotherapeutic techniques. *Amer J. Orthopsychiat.*, 1942, *12*, 429–434.

Counseling and Psychotherapy. Boston: Houghton Mifflin, 1942, 450 pp. Translated into Japanese and published by Sogensha Press, Tokyo, 1951.

1943

Therapy in guidance clinics. *J. Abnorm. Soc. Psychol.*, 1943, *38*, 284–289. Also published in R. Watson (Ed.), *Readings in Clinical Psychology.* New York: Harper and Bros., 1949, 519–527.

1944

Adjustment after Combat. Army Air Forces Flexible Gunnery School, Fort Myers, Florida. Restricted Publication, 1944, 90 pp.

The development of insight in a counseling relationship. *J. Consult. Psychol.*, 1944, *8*, 331–341. Also published in A. H. Brayfield (Ed.), *Readings on Modern Methods of Counseling.* New York: Appleton-Century-Crofts, 1950, 119–132.

The psychological adjustments of discharged service personnel. *Psych. Bulletin*, 1944, *41*, 689–696.

1945

The nondirective method as a technique for social research. *Amer. J. Sociology*, 1945, *50*, 279–283.

Counseling. *Review of Educ. Research*, 1945, *15*, 155–163.

Dealing with individuals in USO. *USO Program Services Bulletin*, 1945.

A counseling viewpoint for the USO worker. *USO Program Services Bulletin*, 1945.

With V. M. Axline. A teacher-therapist deals with a handicapped child. *J. Abnorm. Soc. Psychol.*, 1945, *40*, 119–142.

With R. Dicks and S. B. Wortis. Current trends in counseling, a symposium. *Marriage and Family Living*, 7 (4), 1945.

1946

Psychometric tests and client-centered counseling. *Educ. Psychol. Measmt.*, 1946, *6*, 139–144.

Significant aspects of client-centered therapy. *Amer. Psychologist*, 1946, *1*, 415–422. Translated into Spanish and published in *Rev. Psicol. Gen. Apl.*, Madrid, 1949, *4*, 215–237.

Recent research in nondirective therapy and its implications. *Amer. J. Orthopsychiat.*, 1946, *16*, 581–588.

With G. A. Muench. Counseling of emotional blocking in an aviator. *J. Abnorm. Soc. Psychol.*, 1946, *41*, 207–216.

With J. L. Wallen. *Counseling with Returned Servicemen.* New York: McGraw-Hill, 1946, 159 pp.

1947

Current trends in psychotherapy. In W. Dennis (Ed.), *Current Trends in Psychology*, University of Pittsburgh Press, 1947, 109–137.

Some observations on the organization of personality. *Amer. Psychologist*, 1947, 2, 358–368. Also published in A. Kuenzli (Ed.), *The Phenomenological Problem.* New York: Harper and Bros., 1959, 49–75.

The case of Mary Jane Tilden. In W. U. Snyder (Ed.), *Casebook of Nondirective Counseling.* Boston: Houghton Mifflin, 1947, 129–203.

1948

Research in psychotherapy: Round Table, 1947. *Amer. J. Orthopsychiat.*, 1948, *18*, 96–100.

Dealing with social tensions: A presentation of client-centered counseling as a means of handling interpersonal conflict. New York: Hinds, Hayden and Eldredge, Inc., 1948, 30 pp. Also published in *Pastoral Psychology*, 1952, *3* (28), 14–20; *3* (29), 37–44.

Divergent trends in methods of improving adjustment. *Harvard Educational Review*, 1948, *18*, 209–219. Also in *Pastoral Psychology*, 1950, *1* (8), 11–18.

Some implications of client-centered counseling for college personnel work. *Educ. Psychol. Measmt.*, 1948, *8*, 540–549. Also published in *College and University*, 1948, and in *Registrar's Journal*, 1948.

With B. L. Kell and Helen McNeil. The role of self-understanding in the prediction of behavior. *J. Consult. Psychol.*, 1948, *12*, 174–186.

1949

The attitude and orientation of the counselor in client-centered therapy. *J. Consult. Psychol.*, 1949, *13*, 82–94.

A coordinated research in psychotherapy: A non-objective introduction. *J. Consult. Psychol.*, 1949, *13*, 149–153.

1950

Significance of the self-regarding attitudes and perceptions. In M. L. Reymert (Ed.), *Feelings and Emotions.* New York: McGraw-Hill, 1950. 374–382. Also published in Gorlow, L., and W. Katkovsky (Eds.), *Readings in the Psychology of Adjustment.* New York: McGraw-Hill, 1959.

A current formulation of client-centered therapy. *Social Service Review*, 1950, *24*, 442–450.

What is to be our basic professional relationship? *Annals of Allergy*, 1950, *8*, 234–239. Also published in M. H. Krout (Ed.), *Psychology, Psychiatry, and the Public Interest.* University of Minnesota Press, 1956, 135–145.

With R. Becker. A basic orientation for counseling. *Pastoral Psychology*, 1950, *1* (1), 26–34.

With D. G. Marquis and E. R. Hilgard. ABEPP policies and procedures. *Amer. Psychologist*, *5*, 1950, 407–408.

1951

Where are we going in clinical psychology? *J. Consult. Psychol.*, 1951, *15*, 171–177.

Client-Centered Therapy: Its Current Practice, Implications, and Theory. Boston: Houghton Mifflin, 1951, 560 pp. Also translated into Japanese and published by Iwasaki Shoten Press, 1955.

Perceptual reorganization in client-centered therapy. In R. R. Blake and G. V. Ramsey (Eds.), *Perception: An Approach to Personality.* New York: Ronald Press, 1951, 307–327.

Client-centered therapy: A helping process. *The University of Chicago Round Table*, 1951, *698*, 12–21.

Studies in client-centered psychotherapy III: The case of Mrs. Oak—a research analysis. *Psychol. Serv. Center J.*, 1951, *3*, 47–165. Also published in C. R. Rogers and Rosalind F. Dymond (Eds.), *Psychotherapy and Personality Change.* University of Chicago Press, 1954, 259–348.

Through the eyes of a client. *Pastoral Psychology*, 2 (16), 32–40; (17) 45–50; (18) 26–32. 1951.

With T. Gordon, D. L. Grummon and J. Seeman. Studies in client-centered psychotherapy I: Developing a program of research in psychotherapy. *Psychol. Serv. Center J.*, 1951, *3*, 3–28. Also published in C. R. Rogers and Rosalind F. Dymond (Eds.), *Psychotherapy and Personality Change.* University of Chicago Press, 1954, 12–34.

1952

Communication: Its blocking and facilitation. *Northwestern University Information*, 1952, *20*, 9–15. Reprinted in *ETC*, 1952, *9*, 83–88; in Harvard Bus. Rev., 1952, *30*, 46–50; in *Human Relations for Management*, E. C. Bursk (Ed.). New York: Harper and Bros., 1956, 150–158. French translation in *Hommes et Techniques*, 1959.

A personal formulation of client-centered therapy. *Marriage and Family Living*, 1952, *14*, 341–361. Also published in C. E. Vincent (Ed.), *Readings in Marriage Counseling.* New York: T. Y. Crowell Co., 1957, 392–423.

Client-centered psychotherapy. *Scientific American*, 1952, *187*, 66–74.

With R. H. Segel. *Client-Centered Therapy: Parts I and II.* 16 mm. motion picture with sound. State College, Pa.: Psychological Cinema Register, 1952.

1953

Some directions and end points in therapy. In O. H. Mowrer (Ed.), *Psychotherapy: Theory and Research.* New York: Ronald Press, 1953, 44–68.

A research program in client-centered therapy. *Res. Publ. Ass. Nerv. Ment. Dis.*, 1953, *31*, 106–113.

The interest in the practice of psychotherapy. *Amer. Psychologist*, 1953, *8*, 48–50.

With G. W. Brooks, R. S. Driver, W. V. Merrihue, P. Pigors, and A. J. Rinella. Removing the obstacles to good employee communications. *Management Record*, *15* (1), 1953, 9–11, 32–40.

1954

Becoming a person. Oberlin College Nellie Heldt Lecture Series. Oberlin: Oberlin Printing Co., 1954. 46 pp. Reprinted by the Hogg Foundation for Mental Hygiene, University of Texas, 1956, also in *Pastoral Psychology*, 1956, 7 (61), 9–13, and 1956, 7 (63), 16–26. Also published in S. Doniger (Ed.), *Healing, Human and Divine*. New York: Association Press, 1957, 57–67.

Towards a theory of creativity. *ETC: A Review of General Semantics*, 1954, *11*, 249–260. Also published in H. Anderson (Ed.), *Creativity and Its Cultivation.* New York: Harper and Bros., 69–82.

The case of Mr. Bebb: The analysis of a failure case. In C. R. Rogers, and Rosalind F. Dymond (Eds.), *Psychotherapy and Personality Change.* University of Chicago Press, 1954, 349–409.

Changes in the maturity of behavior as related to therapy. In C. R. Rogers, and Rosalind F. Dymond (Eds.), *Psychotherapy and Personality Change.* University of Chicago Press, 1954, 215–237.

An overview of the research and some questions for the future. In C. R. Rogers, and Rosalind F. Dymond (Eds.), *Psychotherapy and Personality Change.* University of Chicago Press, 1954, 413–434.

With Rosalind F. Dymond (Eds.). *Psychotherapy and Personality Change.* University of Chicago Press, 1954, 447 pp.

1955

A personal view of some issues facing psychologists. *Amer. Psychologist*, 1955, *10*, 247–249.

Personality change in psychotherapy. *The International Journal of Social Psychiatry*, 1955, *1*, 31–41.

Persons or science? A philosophical question. *Amer. Psychologist*, 1955, *10*, 267–278. Also published in *Pastoral Psychology*, 1959, *10* (Nos. 92, 93).

With R. H. Segel. *Psychotherapy Begins: The Case of Mr. Lin.* 16 mm. motion picture with sound. State College, Pa.: Psychological Cinema Register, 1955.

With R. H. Segel. *Psychotherapy in Process: The Case of Miss Mun.* 16 mm. motion picture with sound. State College, Pa.: Psychological Cinema Register, 1955.

1956

Implications of recent advances in the prediction and control of behavior. *Teachers College Record*, 1956, *57*, 316–322. Also published in E. L. Hartley, and R. E. Hartley (Eds.), *Outside Readings in Psychology*. New York: T. Y. Crowell Co., 1957, 3–10. Also published in R. S. Daniel (Ed.), *Contemporary Readings in General Psychology*. Boston: Houghton Mifflin, 1960.

Client-centered therapy: A current view. In F. Fromm-Reichmann, and J. L. Moreno (Eds.), *Progress in Psychotherapy*. New York: Grune & Stratton, 1956, 199–209.

Review of Reinhold Niebuhr's *The Self and the Dramas of History*, *Chicago Theological Seminary Register*, 1956, *46*, 13–14. Also published in *Pastoral Psychology*, 1958, *9*, No. 85, 15–17.

A counseling approach to human problems. *Amer. J. of Nursing*, 1956, *56*, 994–997.

What it means to become a person. In C. E. Moustakas (Ed.), *The Self*. New York: Harper and Bros., 1956, 195–211.

Intellectualized psychotherapy. Review of George Kelly's *The Psychology of Personal Constructs*, *Contemporary Psychology*, *1*, 1956, 357–358.

Some issues concerning the control of human behavior. (Symposium with B. F. Skinner) *Science*, November 1956, *124*, No. 3231, 1057–1066. Also published in L. Gorlow, and W. Katkovsky (Eds.), *Readings in the Psychology of Adjustment*. New York: McGraw-Hill, 1959, 500–522.

With E. J. Shoben, O. H. Mowrer, G. A. Kimble, and J. G. Miller. Behavior theories and a counseling case. *J. Counseling Psychol.*, 1956, *3*, 107–124.

1957

The necessary and sufficient conditions of therapeutic personality change. *J. Consult. Psychol.*, *21*, 1957, 95–103. French translation in *Hommes et Techniques*, 1959.

Personal thoughts on teaching and learning. *Merrill-Palmer Quarterly*, Summer, 1957, *3*, 241–243. Also published in *Improving College and University Teaching*, *6*, 1958, 4–5.

A note on the nature of man. *J. Counseling Psychol.*, 1957, *4*, 199–203. Also published in *Pastoral Psychology*, 1960, *11*, No. 104, 23–26.

Training individuals to engage in the therapeutic process. In C. R. Strother (Ed.), *Psychology and Mental Health*. Washington, D. C.: Amer. Psychological Assn., 1957, 76–92.

A therapist's view of the good life. *The Humanist*, *17*, 1957, 291–300.

1958

A process conception of psychotherapy. *American Psychologist*, 1958, *13*, 142–149.

The characteristics of a helping relationship. *Personnel and Guidance Journal*, 1958, *37*, 6–16.

1959

A theory of therapy, personality, and interpersonal relationships as developed in the client-centered framework. In S. Koch (Ed.), *Psychology: A Study of a Science*, Vol. III. *Formulations of the Person and the Social Context*. New York: McGraw-Hill, 1959, 184–256.

Significant learning: In therapy and in education. *Educational Leadership*, 1959, *16*, 232–242.

A tentative scale for the measurement of process in psychotherapy. In E. Rubinstein (Ed.), *Research in Psychotherapy*. Washington, D. C.: Amer. Psychological Assn., 1959, 96–107.

The essence of psychotherapy: A client-centered view. *Annals of Psychotherapy*, 1959, *1*, 51–57.

The way to do is to be. Review of Rollo May, *et al.*, *Existence: A New Dimension in Psychiatry and Psychology*, in *Contemporary Psychology*, 1959, *4*, 196–198.

Comments on cases in S. Standal and R. Corsini (Eds.), *Critical Incidents in Psychotherapy*. New York: Prentice-Hall, 1959.

Lessons I have learned in counseling with individuals. In W. E. Dugan (Ed.), *Modern School Practices Series 3, Counseling Points of View*. University of Minnesota Press, 1959, 14–26.

With G. Marian Kinget. *Psychotherapie en Menselyke Verhoudingen*. Utrecht: Uitgeverij Het Spectrum, 1959, 302 pp.

With M. Lewis and J. Shlien. Two cases of time-limited client-centered psychotherapy. In A. Burton (Ed.), *Case Studies of Counseling and Psychotherapy*. Prentice-Hall, 1959, 309–352.

1960

Psychotherapy: The Counselor, and *Psychotherapy: The Client*. 16 mm. motion pictures with sound. Distributed by Bureau of Audio-Visual Aids, University of Wisconsin, 1960.

Significant trends in the client-centered orientation. In D. Brower, and L. E. Abt (Eds.), *Progress in Clinical Psychology*, Vol. IV. New York: Grune & Stratton, 1960, 85–99.

With A. Walker, and R. Rablen. Development of a scale to measure process changes in psychotherapy. *J. Clinical Psychol.*, 1960, *16*, 79–85.

1961 (to May 1)

Two divergent trends. In Rollo May (Ed.), *Existential Psychology*. New York: Random House, 1961, 85–93.

The process equation of psychotherapy. *Amer. J. Psychotherapy*, 1961, *15*, 27–45.

A theory of psychotherapy with schizophrenics and a proposal for its empirical investigation. In J. G. Dawson, H. K. Stone, and N. P. Dellis (Eds.), *Psychotherapy with Schizophrenics*. Baton Rouge: Louisiana State University Press, 1961, 3–19.

In Press

Toward becoming a fully functioning person. In A. W. Combs (Ed.), *1962 Yearbook*, Amer. Soc. for Curriculum Development. (In press)

Acknowledgments

Chapter 1, "This is Me," copyright © 1961 by Carl R. Rogers.

Chapter 2, "Some Hypotheses Regarding the Facilitation of Personal Growth," copyright 1954 by Board of Trustees of Oberlin College. Published in pamphlet, "Becoming a person."

Chapter 3, "The Characteristics of a Helping Relationship," copyright 1958 by *Personnel and Guidance Journal.* Published as "Characteristics of a helping relationship," 1958, *37*, 6–16.

Chapter 4, "What We Know About Psychotherapy — Objectively and Subjectively," copyright © 1961 by Carl R. Rogers.

Chapter 5, "Some of the Directions Evident in Therapy," copyright 1953 by Ronald Press. Published as chapter 2, "Some directions and end points in therapy," in O. H. Mowrer (Ed.), *Psychotherapy: Theory and Research*, pp. 44–68.

Chapter 6, "What It Means to Become a Person," copyright 1954 by Board of Trustees of Oberlin College. Published in pamphlet, "Becoming a person."

Chapter 7, "A Process Conception of Psychotherapy," copyright 1958 by American Psychological Association, Inc. Published under the same title in the *American Psychologist*, volume *13*, 142–149.

Chapter 8, "To Be That Self Which One Truly Is: A Therapist's View of Personal Goals," copyright 1960 by Pendle Hill Publications. Published as "A therapist's view of personal goals," Pendle Hill Pamphlet #108.

Chapter 9, "A Therapist's View of the Good Life: The Fully Functioning Person," copyright 1957 by *The Humanist*, Humanist House,

Yellow Springs, Ohio. Published as "A therapist's view of the good life," volume *17*, 291–300.

Chapter 10, "Persons or Science? A Philosophical Question," copyright 1955 by the American Psychological Association, Inc. Published under same title in the *American Psychologist*, volume *10*, 267–278.

Chapter 11, "Personality Change in Psychotherapy," copyright 1955 by *The International Journal of Social Psychiatry*. Published under same title in volume *1*, 31–41.

Chapter 12, "Client-Centered Therapy in Its Context of Research," copyright 1959 by Uitgeverij Het Spectrum, Utrecht, The Netherlands. Published as chapter 10 in *Psychotherapie en Menselyke Verhoudingen*, by C. R. Rogers and G. M. Kinget.

Chapter 13, "Personal Thoughts on Teaching and Learning," copyright 1957 by *Merrill-Palmer Quarterly*. Published under same title in volume *3*, 241–243.

Chapter 14, "Significant Learning: In Therapy and in Education," copyright 1959 by *Educational Leadership*. Published under same title in volume *16*, 232–242.

Chapter 15, "Student-Centered Teaching as Experienced by a Participant," copyright 1959 by *Educational Leadership*. Published under title "Carl R. Rogers and nondirective teaching," volume *16*, February, 1959.

Chapter 16, "The Implications of Client-Centered Therapy for Family Life," copyright © 1961 by Carl R. Rogers.

Chapter 17, "Dealing with Breakdowns in Communication — Interpersonal and Intergroup," copyright 1952 by *ETC: A Review of General Semantics*. Published under title "Communication: its blocking and facilitation" in volume *9*, 83–88.

Chapter 18, "A Tentative Formulation of a General Law of Interpersonal Relationships," copyright © 1961 by Carl R. Rogers.

Chapter 19, "Toward a Theory of Creativity," copyright 1954 by *ETC: A Review of General Semantics*. Published under same title in volume *11*, 249–260.

Chapter 20, "The Growing Power of the Behavioral Sciences," copyright © 1961 by Carl R. Rogers.

Chapter 21, "The Place of the Individual in the New World of the Behavioral Sciences," copyright © 1961 by Carl R. Rogers.

Index

* Italicized numbers indicate a bibliographic reference.